E.T. Barnum

E. T. Barnum, Manufacturer of Bank Railings

Iron Fencing, Roof Cresting, Wire and Iron Work

E.T. Barnum

E. T. Barnum, Manufacturer of Bank Railings
Iron Fencing, Roof Cresting, Wire and Iron Work

ISBN/EAN: 9783337121969

Printed in Europe, USA, Canada, Australia, Japan

Cover: Foto ©berggeist007 / pixelio.de

More available books at **www.hansebooks.com**

LATEST PATTERNS

—OF—

IMPROVED WIRE FLOWER POT STANDS.

I offer to the trade the largest variety of goods in my line made by any one manufacturer, and at very low prices. I desire to call especial attention to the reduced prices of my **Wire Flower Stands.** Their SUPERIOR STRENGTH and EXTRA FINISH have always given them a preference over those of other manufacturers. The prices at which I am now offering them is as low as any first class Flower Stand can be sold.

They are all painted a beautiful green, and bronzed with gold bronze, making them very attractive and ornamental.

They can be shipped as freight to all parts of the country, with but trifling expense, although three or four Stands can often be sent a long distance as cheaply as one. I am shipping these goods to all the principal cities in the United States.

No. 5 Stand, with Arch and Basket, $8.00.

ONE OF THE BEST SELLING STANDS IN THE MARKET.

Stand and Arch, 6 feet 4 inches high; length, 4 feet; depth, 32 inches; top shelf, 12 inches diameter. The Arch is convenient for training plants and vines, making it very attractive in a front or bay window.

No. 5½ Gothic Arch Stand, Bronzed, $9.00.

This Stand is universally admired. Stand and Arch, 6 feet 7 inches high; width, 4 feet; depth, 32 inches; top shelf, 12 inches diameter.

No. 1 Stand, Bronzed, $6.50.

[illegible] Stand, and occupies a very little space. Length of shelves 3 [illegible] high, 4 feet wide and 32 inches deep, [illegible] to 30 pots.

Nos. 20 and 21 are [illegible] as well as the window.

No. 2 Square Stand, $7.50.

Size, 40 inches high, 20 inches deep; shelves 39 inches long; will [illegible] 20 pots. This Stand is desirable for piazzas. [illegible] ple Stand, [illegible]

No. 10 Stand, with Arch and Basket, $10.50.

A New Pattern, and the Handsomest Stand in the Market.

Stand and Arch, 6 feet 8 inches high; length, 4 feet; depth, 32 inches. This is a large and very attractive Stand, and unsurpassed for strength and ornamental appearance.

No. 17 Gothic Arch Stand, $13.00.

7 feet high; shelves 40 inches long, 17 inches wide.

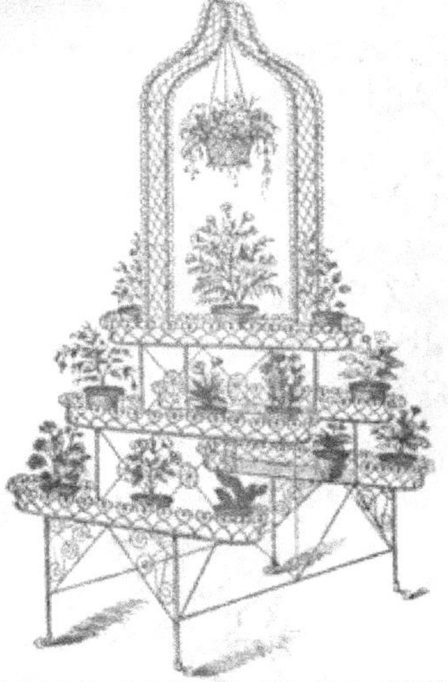

No. 20 Double Stand, with Arch, $11.00.

Stand and Arch is 6 feet 7 inches high; Stand without Arch, 4 feet 6 inches high. The depth is 3 feet 10 inches. Length of shelves 3 feet 3 inches.

No. 21 Square Stand, with Arch, $8.50.

Stand and Arch 6 feet 7 inches high; without Arch, 3 feet 6 inches high, 30 inches deep, and shelves 3 feet 3 inches long.

Nos. 20 and 21 are both the same shape, except the No. 20 is a Double Stand, which gives a front for the room as well as the window.

No. 11 Stand, new style, $8.50.

This is a new pattern, is large in size, and will hold about 40 flower pots. Size, 3 feet 7 inches high; length, 4 feet; depth, 32 inches.

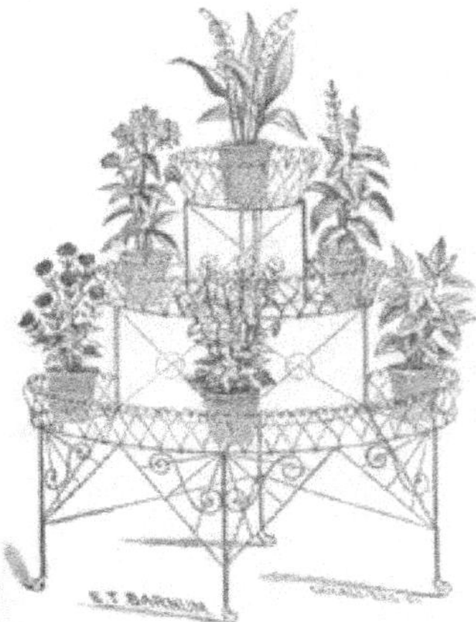

No. 9 Triangular Corner Stand, $6.00.

Size, 3 feet 5 inches high, and three-cornered, so as to fit in the corner of a room, making it very attractive.

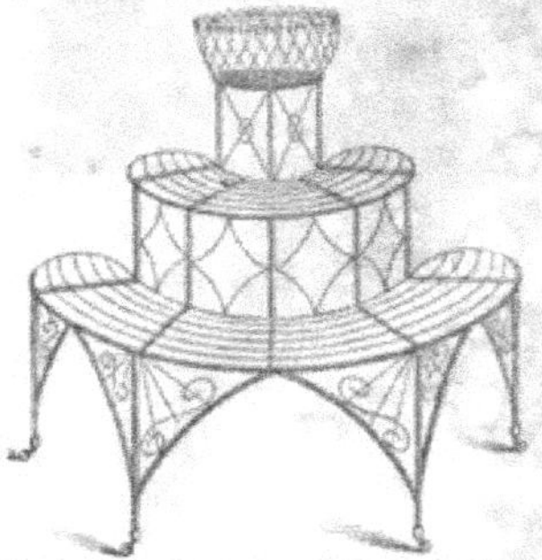

No. 6 Stand, without Border, $5.50.

3 feet 5 inches high, 4 feet long and 32 inches deep. This is made without a border, so as to receive extra large heavy pots. It is made with castors, and can be moved to any part of the room with ease.

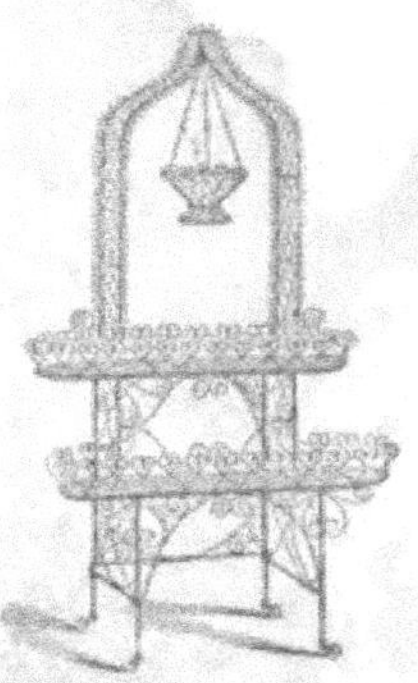

No. 8 Square Stand, with Arch, $6.50.

This is a very attractive Stand, and occupies very little space. Stand and Arch, 5 feet 7 inches high, and 18 inches deep; shelves are 32 inches long.

No. 19 Stand, with Arch, $5.00.
" 19 Stand, with Pan, 6.00.

Stand and Arch, 5 feet high; shelf, 32 inches long, and 12 inches wide. A water-tight pan is fitted in at the base for sand, moss and flowers.

No. 8½ Square Stand, $4.50.

Size, 31 inches high, 18 inches deep; shelves, 33 inches long; it holds a large number of pots for the room it occupies.

No. 24 Revolving Stand, $8.00.

This Stand is 3 feet 8 inches high, and 28 inches diameter. It is mounted on an ornamental pedestal, the same as No. 23. The shelves are high and revolve, which is a great convenience in watering.

No. 27 Revolving Stand, Bronzed, $14.00.

Size, 4 feet 6 inches high, and 42 inches diameter; will hold about 25 pots. It is mounted on an ornamental pedestal with four feet and porcelain castors. To keep the plants in a thriving condition, all of them should have an equal share of the light and warmth of the sun. With this Revolving Stand, the plants can always be kept in a healthy condition.

No. 23 Revolving Stand, $7.00

Size, 3 feet 6 inches high and 24 inches diameter; mounted on a cast iron pedestal with three feet and porcelain castors. It is a very attractive Stand for a small room, and convenient for plants that require plenty of light and sun.

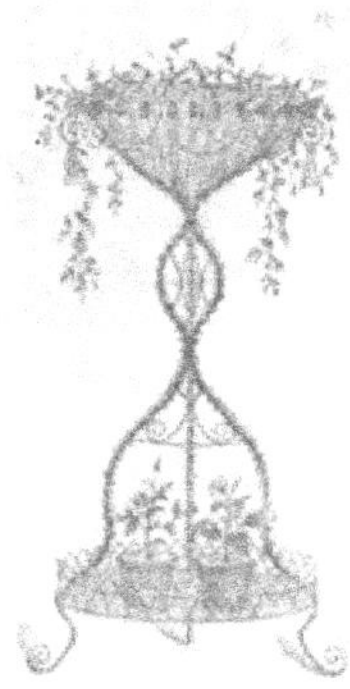

No. 13 Moss, Flower and Bouquet Stand.

26 to 30 inches high. . . . . .each, $3.50

No. 14 Fruit or Plant Stand, $10.00.

This is 4 feet 6 inches high, and has three revolving shelves and a basket. It is used as a Fruit Stand for displaying oranges, lemons, apples, &c.

No. 4½ Extra Large Stand, Bronzed, $14.00.

Stand and Arch, 6 feet 10 inches high; without Arch, 42 inches high; width, 4 feet 8 inches, and depth, 3 feet 4 inches; top shelf, 15 inches diameter. This is an extra large Stand, and is suitable for large bay windows, conservatories, &c., and will hold a great many pots.

No. 29 Round Circle Stand, $12.00.

Stand and Arch 6 feet 7 inches high; without Arch, 3 feet 6 inches high; diameter, 4 feet. This Stand is made with the shelves running all the way round (the cut only shows the shelves part of the way). Can be used for plants, oranges, fruit, &c.

No. 25 Three Legged Stand, $6.00.

3 feet high, 42 inches wide, 26 inches deep.

No. 41 Stand, $5.50.

Size, 33 inches high, 20 inches deep; shelves 36 inches long.

No. 31 Wire Guard for Window Shelf.

2 feet [illegible] inches long, 4½ inches high, 6 inches wide, each $2.00
[illegible] 4½ " " 7 " " " 2.50
[illegible] 4½ " " 7 " " " 3.00

[illegible] MADE TO ORDER.

[illegible] intended to be put on the outside of the window frame [illegible] pots, to keep them from falling out. The [illegible] are made of wire, with a fastening at the [illegible] window frame, or they can be used inside, [illegible] made with galvanized pan, 75 [illegible]

No. 39 Stand.

12 inches high, 24 inches long, 10 inches wide $2.50
13 " 28 " 11 " 3.00
14 " 32 " 12 " 3.25

This Stand is intended to set in front of a low window; can be made with galvanized iron pan for $1 extra.

No. 25½, with Arch, $7.50.

5 feet 7 inches high, 42 inches wide, and 26 inches deep This is a low priced Stand, and very ornamental.

No. 26 Flower Pot Stand, $12.00.

Stand and Arch, 6 feet 7 inches high; Stand, without Arch, 3 feet 5 inches high; depth, 30 inches; length, 3 feet 10 inches. When filled with flower pots, this is one of the most attractive flower stands made. It is handsomely painted and bronzed, and highly ornamental.

No. 40 Square Stand, $7.00.

This is a very ornamental Stand, and occupies but little room. Stand and Arch, 5 feet 9 inches high. Without Arch, 34 inches. Shelves are 36 inches long, and 20 inches deep.

No. 33 Flower Bed Border and Arch.

4 feet diameter, border 9 in. high, Arch [illegible]
5 " " " 12 " [illegible]
6 " " " 14 " [illegible]

Border only, 12 inches [illegible] painted, ready for setting [illegible]

No. 32 Jardiniere and Flower Bed Stand.

20 inches high, 8½ inches wide, and 30 inches long ... $12.00
34 " 9 " 32 " ... 15.00

These Stands are intended for conservatories and bay windows, and to take the place of the expensive tile boxes. The box is made of walnut, lined with zinc, with a heavy ornamental moulding around the top and base, polished and varnished. The gilt stand upon which the box rests, is gilded all over with 23 carat gold, making a very handsome and attractive bay window ornament. The stand can be painted or bronzed (not gilded) costing $3 less.

No. 36 Aquarium Flower Stand.

Size bottom shelf, 3 feet long, 2 feet wide.... ....$18.00

This combination of an ornamental Wire Stand and Aquarium makes a very attractive appearance. The plants being near to water, do well, and become very thrifty in growth. The stand is handsomely painted and bronzed. Any size made to order. Can be made with arch if desired.

No. 35 Lambrequin Trainer.

This is a new and novel way of training vines, and presents a handsome appearance, both from an interior and exterior view of the room. They are made to order, any size that may be required, ranging in price from $5 to $10 each, according to size and quantity.

No. 34 Lyre Trainer.

4 feet high, on castors ... $6.50
5 " " ... 7.50
6 " " ... 8.50

This is a new and desirable stand for ivy or any training vines, and very ornamental.

Wire Foundations and Emblems

TO USE WITH EVERGREENS IN MAKING

CHURCH OR SCHOOL DECORATIONS.

IT GREATLY IMPROVES THE WORK AND CAN BE USED YEAR AFTER YEAR.

THE PATENT BODY OR FOUNDATION.

That it may be convenient for general use, is put up in bundles of 100 feet.

PRICE BY THE QUANTITY.

No. 1	$3.50	per hundred feet net.
" 2	2.50	" " "

To strangers my terms are Cash with the order, or satisfactory references.

This work is made of a continuous metal strip with a circle or spiral wire, for holding the evergreens, forming a strong, solid and continuous strip for festooning or decorating which cannot pull apart. It can be ordered in any lengths desired. I also make the frame work of the same material for a large variety of designs, a few of which are given below. Parties requiring these goods will please send for special catalogue of other styles and price list.

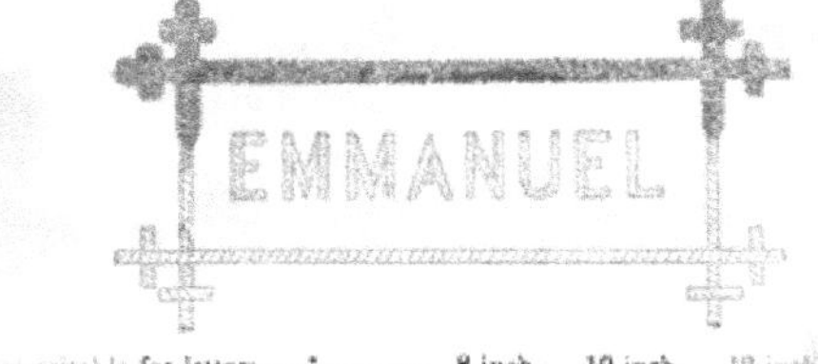

Letters or Figures.

Frames suitable for letters	8 inch.	10 inch.	12 inch.
The length of frame, inside	6⅝ feet.	7⅜ feet.	7¾ feet.
Price	$1.25	$1.40	$1.55

6 in. each,	$.08	12 in. each,	$.15
8 "	.10	15 "	.20
10 "	.12		

Only list price will be charged for the letters in addition. They will be secured to a wire to hang in frames if desired, but where they can be attached to the wall with brads, the wire is not necessary.

LARGE BANNERS FOR EVERGREENS.

FLOWER POT BRACKETS.

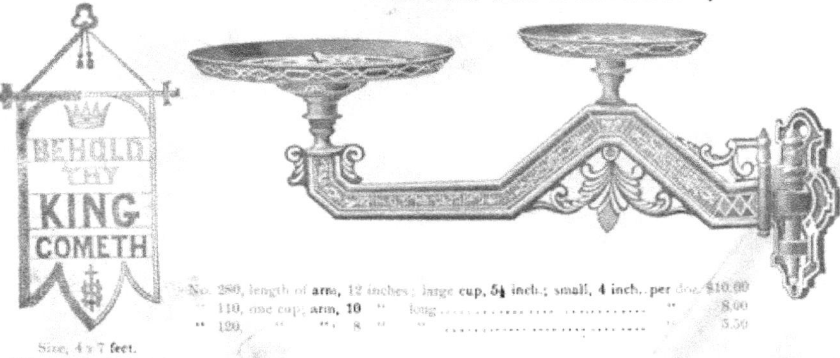

No. 280, length of arm, 12 inches; large cup, 5½ inch.; small, 4 inch..per doz. $10.00
" 110, one cup; arm, 10 " long .... " 8.00
" 120, " ", 8 " " .... " 5.50

Size, 4 x 7 feet.
Price .... $5.50 net.

(2c)

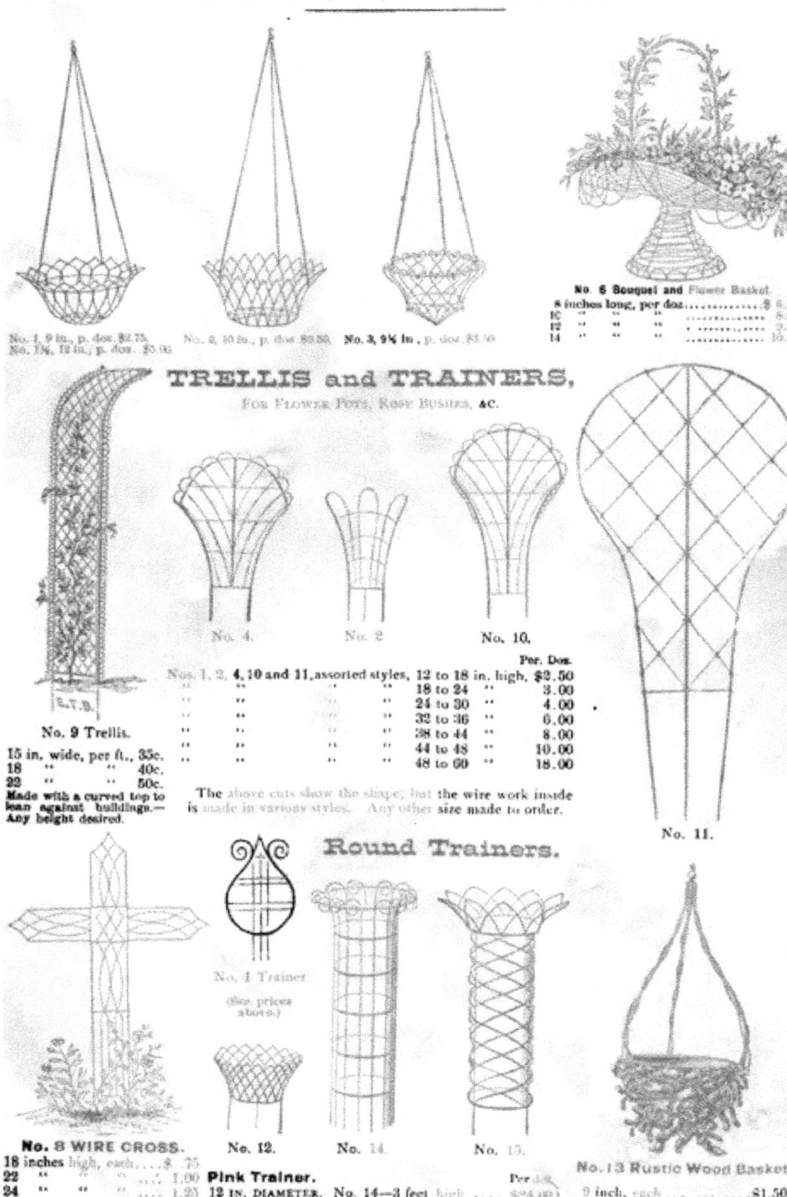
No. 1, 9 in., p. doz. $2.75.
No. 1½, 12 in., p. doz. $5.00.
No. 2, 10 in., p. doz.
No. 3, 9½ in., p. doz.
No. 6 Bouquet and Flower Basket.
8 inches long, per doz.....$ 6.00
10 " " " 8.00
12 " " " 9.00
14 " " " 10.00
TRELLIS and TRAINERS,
FOR FLOWER POTS, ROSE BUSHES, &C.
No. 4.
No. 2
No. 10.
Per. Doz.
Nos. 1, 2, 4, 10 and 11, assorted styles, 12 to 18 in. high, $2.50
" " " " 18 to 24 " 3.00
" " " " 24 to 30 " 4.00
" " " " 32 to 36 " 6.00
" " " " 38 to 44 " 8.00
" " " " 44 to 48 " 10.00
" " " " 48 to 60 " 18.00
The above cuts show the shape; but the wire work inside is made in various styles. Any other size made to order.
E.T.B.
No. 9 Trellis.
15 in. wide, per ft., 35c.
18 " " 40c.
22 " " 50c.
Made with a curved top to lean against buildings.—Any height desired.
No. 11.
Round Trainers.
No. 1 Trainer
(See prices above.)
No. 8 WIRE CROSS.
18 inches high, each.....$ .75
22 " " " 1.00
24 " " " 1.25
30 " " " 1.50
36 " " " 2.50
No. 12.
No. 14.
No. 15.
Pink Trainer.
12 IN. DIAMETER.
Per doz.....$6.00
Per doz.
No. 14—3 feet high.....$24.00
" 15—3 " .....
No. 13 Rustic Wood Basket
9 inch, each.....$1.50
10 " " 1.75
11 " " 2.00

ARCHWAYS AND ARBORS,

FOR GARDENS, DOORWAYS, WALKS, ENTRANCE GATES, ETC.

No. 6 Gothic Top, Double Border.

7 feet high,	4 feet span,	2 feet wide	........each	$14.00
8 "	5 "	2 "	.... "	16.00
9 "	6 "	2 "	.... "	18.00

Arbor For Climbing Shrubs and Vines.

No. 7,	4 feet high,	24 inch span,	12 inches wide	each	$4.00
" 8,	4½ "	26 "	13 "	"	5.00
" 9,	5 "	2[illegible] "	14 "	"	6.00
" 10,	6 "	36 "	16 "	"	8.00

WIRE GRAVE GUARDS,

Designed either for Grown Persons' or Children's Graves. Any size or style made to order, painted green or drab. Also, name can be put on any of the different styles.

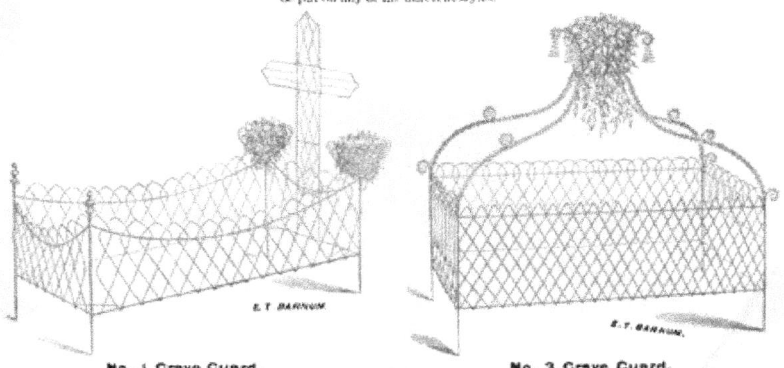

No. 1 Grave Guard.

3 feet long,	18 inches wide	........each	$6.00
3½ "	21 "	.... "	7.50
5 "	24 "	.... "	9.00

Without the Cross and Baskets, one dollar less.

No. 3 Grave Guard.

3 feet long,	18 inches wide	........each	$6.50
3½ "	21 "	.... "	8.50
5 "	24 "	.... "	10.00
6 "	30 "	.... "	12.00

These Guards or Railings are very generally used in Cemeteries, being very appropriate for ornamenting graves. They are durable, and one coat of paint in two years will make them as fresh and bright as new.

The name or initials, put on either end, will cost 25 cents per letter for full name, or $1 extra for two or three initials.

New Summer House.

No. 1, 7 feet deep, 6 feet wide, 8 feet high .... each $50.00
" 2, 8 " 7 " 9 " .... " 60.00

This is the most attractive summer house for the price now in the market. When covered with foliage and climbing vines they are a delightful resort in a hot summer day, cool and airy, as the wire work has no surface to obstruct even a light breeze. Seats are usually furnished by the purchaser and arranged to suit his own taste, and are not included in above prices. They are painted a beautiful green, or any other color if preferred, and shipped without difficulty as freight to all parts of the country.

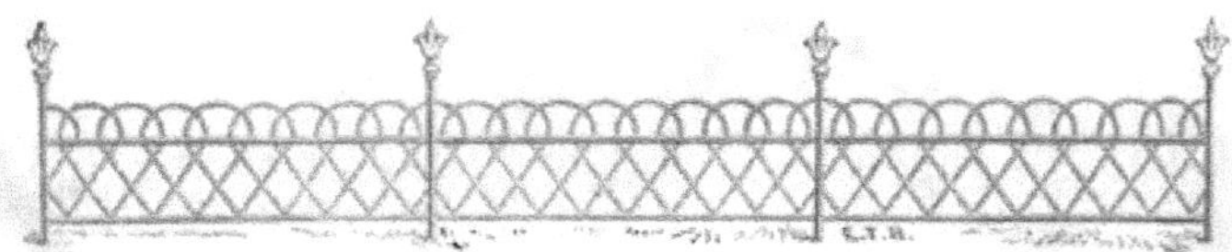

No. 32, Wire Fence or Lawn Bordering, 9 inches high .... per foot, 30c
" 32, " " 12 " .... " 35c

Made 3½ inch crimped, diamond mesh wire work, of Nos. 9 and 10 wire. The pickets or posts are ⅜ inch iron with ornaments on top. The border is made in any lengths from 10 to 60 feet, and rolled up for shipping. It can be unrolled and cut off any length, or bent any shape desired. This wire bordering is suitable for any purpose requiring a light inexpensive railing. It can be made with feet to fasten to wood if desired.

Open Summer House.

No. 4, 7 feet deep, 6 feet wide, 8 feet high ....each $40.00
" 5, 8 " 7 " 9 " .... " 50.00

This style is very desirable for locating in the centre of grounds with a continuous walk, and when covered with foliage or climbing vines is a very attractive and pleasant resort. Seats are usually furnished by the purchaser.

Other sizes than those given above can be made if desired.

Iron Lawn Border.

No. 31, 12 inches high ....per foot, 45c

Fastens together in sections with projecting points to set in the ground.

WIRE SUMMER HOUSES.

No. 8, 8 feet diameter, 10 feet high..............................each $100.00
" 10, 10 " 12 " .... " 125.00

Made octagon in shape with a strong iron **frame, and put together in sections as shown in cut. Painted a handsome green or any other color** if preferred. Seats **are not included in above prices, but are usually furnished by the purchaser. An opening can be made** at back same **as front without additional charge. Shipped as freight, in sections, to all** parts **of the country.**

Wire Spark Guard, French Top.

No. 32, 25 inches high, 23 inches wide, 7½ inches deep..............................each $3.50
" 33, 27 " 24 " 8 " .... " 4.00
" 34, 28 " 25 " 8½ " .... " 4.50

Made of plated wire **and so** as to fit closely any **coal** or wood grate. It hooks on to the grate and is easily taken **off and put on,** and prevents sparks from flying **out. It is not** safe to leave a fire in a grate without this protection. If **desired extra fine,** add $1.00 to above prices; **can be made all brass** wire at twice the above prices. If made other sizes **a small additional** price will be charged on all **guards. In ordering** spark guards the measure **is taken** including **a lap of about 2 inches** all around.

Nursery Fenders and Spark Guards.

To put in front of grates or fire places and around stoves and heaters. The meshes are fine enough to protect the carpet from sparks and dresses and children from coming in contact with the fire. The usual sizes are from 34 to 38 inches long and from 20 to 30 inches high.

Improved Wire Nursery Fenders.

WITH POLISHED BRASS TOP.

No.	Length	Height	Depth		Price
No. 21,	32 inches long,	16 inches high,	14 inches deep,	bronzed..each	$5.00
" 22,	34 "	18 "	15 "	" .. "	5.50
" 23,	36 "	20 "	15 "	" .. "	6.50
" 24,	38 "	24 "	15 "	" .. "	7.50
" 25,	40 "	30 "	16 "	" .. "	8.50

Other sizes made to order at a small additional price.

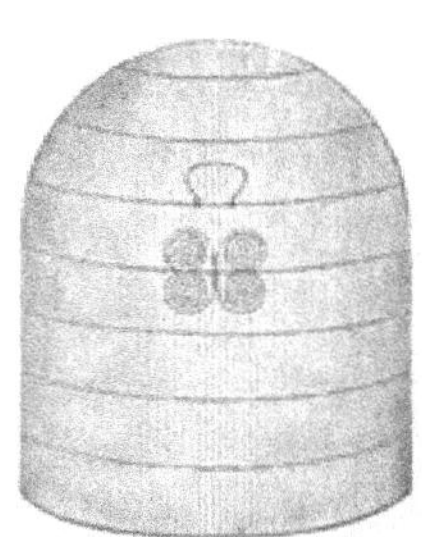

Wire Spark Guard, Round Top.

No.	High	Wide	Deep	Price
No. 35,	26 in. high,	20½ in. wide,	7½ deep,	$3.00
" 36,	27 "	21 "	8½ "	3.50
" 37,	27 "	22 "	9 "	4.00
" 38,	28 "	24 "	9½ "	4.80

Is made of diamond meshes, plated wire, and so as to fit closely any grate. It is easily taken off and put on, and prevents sparks from flying out. Can be made to order any size desired.

If made to order other sizes than above a small additional price will be charged.

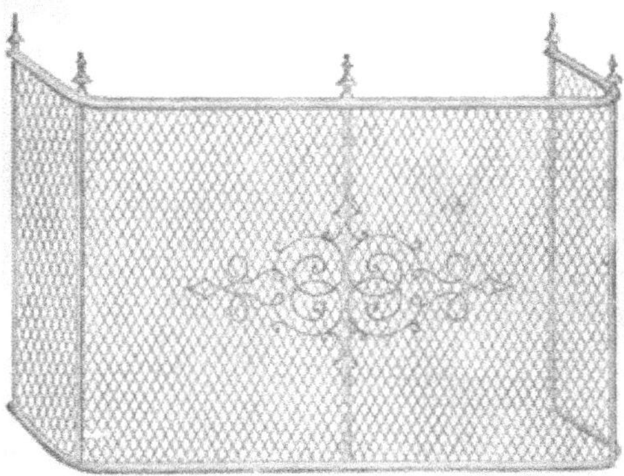

Improved Wire Nursery Fenders.

WITH POLISHED BRASS TOP AND ORNAMENTS.

No.	Length	Height	Depth		Price
No. 27,	34 inches long,	16 inches high,	14 inches deep,	bronzed, each	$6.50
" 28,	36 "	20 "	15 "	" "	7.50
" 29,	38 "	24 "	15 "	" "	8.50
" 30,	40 "	30 "	16 "	" "	10.00

This Fender is a very attractive parlor ornament, can be made all polished brass with a handsome O G brass moulding at bottom for twice the above prices. If made other sizes than above a small additional price will be charged. If made with a plain oval bronzed top, $2.00 less.

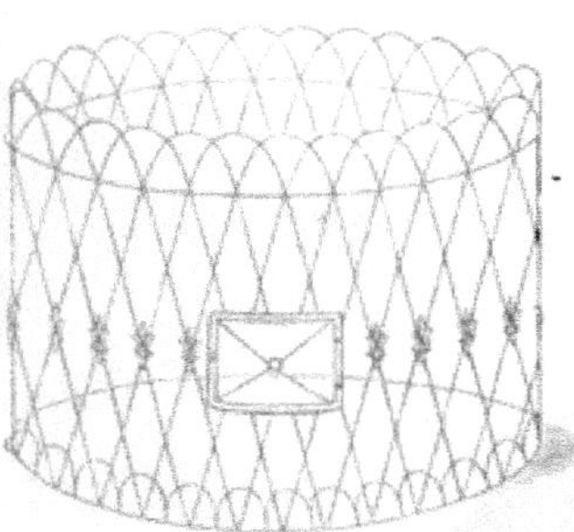

No. 5 Wire Stove Guard.

MADE TO ORDER ANY SIZE.

2 inch plain diamond..per square foot, 25c
With rosettes.......... " " 30c

Where there are children, this is an indispensable protection to put around stoves, heaters, etc. It can be made with a door to open all he way up, or with a small door, same as cut. In ordering give the diameter, height, and the height to have the door from the floor.

No. 13 Chair, Grape Pattern, $8.00
To match No. 10 Settee.

No. 10 Iron Settee, Grape Pattern.

No. 10, 2 seat Grape Settee, bronzed	....	$12.00
" 11, 3 " " " "	....	14.00
" 12, 4 " " " "	....	16.00

No. 9 Rustic Iron Chair, $8.00

No. 8 Rustic Iron Settee, 4 seat $14.50

No. 17 Heavy Wire Arm Chair, $6.50

No. 24 Fern Chair, Bronzed, $12.00
New pattern, cast iron, representing fern leaves.

All my Settees and Chairs are handsomely painted with metalic paint and decorated with gold bronze.

No. 25 Fern Leaf Settee, 4 seat, $18.00.

This is a new pattern, representing fern leaves, and is one of the richest cast iron Settees made. Intended especially for large, elegant lawns and parks. Handsomely painted and finished with gold bronze. No. 24 Chair is exactly the same pattern.

HEAVY WIRE CHAIRS AND SETTEES.

No. 18 Heavy Wire Settee, 3 seat, $15.00. **No. 1 Wire Chair, $5.00.**

They are very comfortable and suitable for lawns, piazzas, offices and cemetery lots. Wire Chairs or Settees are stronger than wood or iron, and much lighter to handle. They are handsomely painted and bronzed.

Philadelphia Lawn Mower.

THE BEST IN THE WORLD.

	Width of Cut.		Power Required.	Weight.	Price.
No. 00.	10 inch,	6½ inch wheels,	A Lady,	32 lbs.	$14.00
" 0.	12 "	6½ "	"	34 "	18.00
" 1.	14 "	6½ "	A Youth.	37 "	20.00
" 2.	16 "	6½ "	One Man.	41 "	22.00
" 3.	18 "	6½ "	"	46 "	24.00
" 4.	20 "	6½ "	"	50 "	26.00
" 5.	15 "	8½ "	"	51 "	22.00

Cuts as well *turning a short corner* as when pushed straight forward.

With proper care this Mower will last twenty years, which reduces the yearly expense for cutting the grass **to a trifle over** one dollar, making it a paying investment. Every one warranted to give satisfaction.

The Philadelphia Lawn Mower has several important parts of the machine patented which are very essential and indispensable to make a mower perfect, and it is not a matter of surprise, therefore, that others are endeavoring **to sell** worthless machines on the reputation of the Philadelphia Mower. Extra parts of machines kept in stock.

Horse Machines—New Pattern.

To cut 30 inch,	with Draft Rod,	Light Horse, 315 lbs.	$100.00
" 30 "	Shafts and Seat,	" " 350	120.00

IRON VASES.

Fluted Vase.

No. 1, 16 in. high	$ 4.00;	same, 28 in. high, with	Pedestal	$ 7.00
" 2, 20 "	6.00;	" 38 "	"	10.00
" 3, 24 "	8.00;	" 44 "	"	14.00
" 4, 30 "	12.00;	" 54 "	"	20.00
" 5, 36 "	25.00;	" 63 "	"	37.00

Venitian Vase.

40 inches high, 21 inches diameter $17.00

Send for special catalogue of Vases and Fountains.

Lawn, Garden and Cemetery Ornaments.

DEER RECLINING.

Deer reclining, 3 feet 2 inches long, painted one coat	$65.00
" 3 " 2 " natural color	67.50
Deer standing, 3 " 4 inches, 4 feet long, painted one coat	75.00
" 3 " 4 " 4 " natural color	80.00

No. 14 Reservoir Vase.

No. 13, 20 in. high, Vase only, no Pedestal	$10.00
" 14, 35 " as shown in cut	14.00

Diameter of vase at top 23 in., base 14 in. square.

No. 17 Reservoir Vase.

No. 16, 27 in. high, Vase only, no Pedestal	$10.00
" 17, 41 " as shown in cut	15.00

Diameter of vase at top, 22 in.; base, 14 in. square.

☞ Send for Special Catalogue of Vases, Fountains, &c.

No. 205 Shield and Leaf Vase.

60 inches high	each	$48.00
72 "	"	87.00
82 "	"	115.00

Diameter of Vase, including handles, 42 inches. Diameter of Vase, without handles, 32 inches.

This is the handsomest vase for the price in the market.

SPITZ DOG.

No. 6, 2 feet long, 17 inches high, painted white $25.00
" 6, 2 " " " " natural color 27.50

No. 269, Cupid and Fish Fountain, 3 feet high, painted one coat, without basin $50.00
With 4 feet diameter Frog and Turtle Ground basin, complete 75.00

☞ Send for Catalogue of other styles of fountains.

Lawn Tent and Settee.

No. 32, 4 feet 3 inches long .... each $25.00
" 34, 5 " long .... " 50.00

This is the most convenient Lawn Tent in use. The canopy can be folded up or thrown backward or forward as may be desired. The Settee is hard wood and all can be conveniently packed for shipment.

Lawn or Park Settee.

AQUARIUMS.

No. 14 Aquarium.

New and very neat; gotten up with a view to meeting the demand for a "TASTEFUL LOW PRICED AQUARIUM."

Size of base, 13 x 18 inches $30.00

No. 16 Aquarium.

No. 16, 26 inches long, 16 in. wide and 12 in. high, glazed $25.00
" [illegible] 14 " 14 " " 18.00

Larger sizes made to **order**.

CRIMPED WIRE WINDOW GUARDS.

Prices of Nos. 1, 3 and 5 Crimped Window Guards:

Mesh	Wire	Frame	Per sq. ft.
¾ inch mesh	No. 16 wire	frame No. 12	.30
1 "	" 14 "	" 11	.25
1 "	" 13 "	" 10	.30
1¼ "	" 14 "	" 10	.25
1¼ "	" 13 "	" 9	.28
1¼ "	" 12 "	" 9	.30
1½ "	" 13 "	" 10	.25
1½ "	" 12 "	" 9	.30
1½ "	" 11 "	" 8	.35
1½ "	" 11 "	" 8	.30
1¾ "	" 10 "	" 7	.35
2 "	" 10 "	" 7	.35
2 "	" 9 "	" 7	.38
2 "	" 8 "	" 6	.45

No. 1 Wire Window Guard

For Churches or Public Buildings, WITH ARCH TOP.

No. 3 Crimped Window Guard.

Showing a Finished or Scroll Top, suitable for Store Windows, Doors, etc.

No. 5 Crimped Window Guard.

For Basement Windows of Hotels, Private Residences, etc.

I present above illustrations of the most common patterns, but make any style required. 1¼ inch mesh, No. 13 wire, is generally used for church, school house, barn and cellar windows, and painted green or drab. Special price for large orders. They should lap over about ¼ inch on the frame, for stapling on. They do not obstruct the light, are strong and durable, and a sure protection against burglars.

☞ In ordering, state which is height and give exact size you want them made, including the lap of ¼ inch or more as above.

No. 6 Crimped Wire Guard with Picket Top.

This guard has a round iron frame to fasten with staples (or screws if desired). The top is channel iron which greatly strengthens the guard. Suitable for doorways and for the lower part of large or small windows, etc.

1¼ inch mesh, No. 12 wire................................per sq. foot, .35
1½ " " 11 "................................" " [illegible]
2 " " 10 "................................" " .40
2½ " " 8 "................................" " [illegible]

This class of wire work is also used for enclosing elevators, partitions for offices, &c.

WROUGHT IRON WINDOW GUARDS.

No. 15 Pattern Heavy Wrought Iron Wire Work for Vault Doors.

This style of heavy wire work is very strong and ornamental, and is used for doors in Cemetery Vaults, also in Banking Offices and Stores. It can be made any size with plain or arched top and with such locks and fastenings as the location requires.

No. 15, as cut, made with ¼ inch wire, per sq. foot, $1.40
" 15, " " 5-16 " " 1.50

If made with arched or gothic top, $2.00 extra.

Locks and fastenings extra, $[illegible]

No. 7 Heavy Ornamental Guard with Channel Iron Frame.

2½ inch mesh, No. 6 (3-16 inch) wire..per sq. foot, $1.00
2 " " 4 ¼ " ".. " " 1.15

If made all plain diamond mesh, deduct 15c. per sq. foot from above prices. The frame is made of channel iron to set into a jam and fasten with screws. See cut No. 21, page 28, representing method of making this work. In ordering, state which is height and give exact size wanted. Where these guards are used for store fronts, goods in show windows can be displayed at night in safety. They are made to order all sizes and shapes desired.

HEAVY CRIMPED WIRE WINDOW GUARDS.

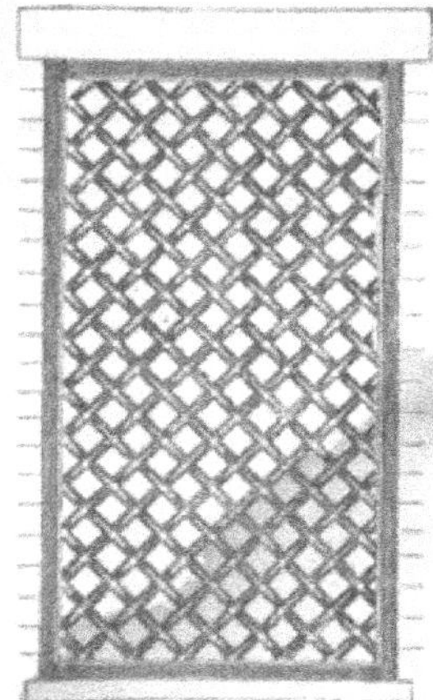

No. 8 Extra Heavy Prison or Asylum Guard.

(WITH A CHANNEL IRON FRAME.)

This is an extra heavy Crimped Steel Wire Guard, made with a channel iron frame and used for prisons, insane asylums and other purposes, where an extra strong guard is required. In ordering, state if it is to screw on the outside or set in the inside of the jam. If on the outside frame will be made of round iron and fasten with staples. See cuts No. 20 and 21 on page 28 for both styles of frames above referred to.

¾ inch mesh, No. 10 wire .... per sq. foot, $ .75
1 " " 9 " .... " " .70
1¼ " " 7 " .... " " .80

No. 4 Heavy Crimped Diamond Guard.

(WITH A CHANNEL IRON FRAME.)

1¼ inch mesh, No. 11 wire (⅛ inch) .... per sq. foot, $ .40
1½ " " 10 " (3-32 ") .... " " .50
2 " " 9 " .... " " .60
2 " " 8 " .... " " .65
2½ " " 7 " (3-16 ") .... " " .75

These guards fasten securely to the window frame with screws, the frame being set into a jam. See cut No. 21, page 28, for channel iron frame.

No. 9 Extra Heavy Crimped Guard.

1¼ inch mesh, No. 9 wire .... per sq. foot, $ .55
1½ " " 8 " .... " " .60
2 " " 7 " .... " " .65
2½ " " 6 " .... " " .75

The above guards are made to order any size desired, are suitable for public buildings or private residences.

No. 10 Extra Heavy Crimped Guard.

1¼ inch mesh, No. 9 wire .... per sq. foot, $ .60
1½ " " 8 " .... " " .65
2 " " 7 " .... " " .70
2½ " " 6 " .... " " .90

The above guards often take the place of cast iron, being cheaper, and much stronger and more ornamental and durable.

The above prices are for ordinary size windows.

No. 20.

Cut showing method of making Nos. 1, 3, 5 and 6 Crimped Wire Window Guards with round iron frame and how the wire is secured to it.

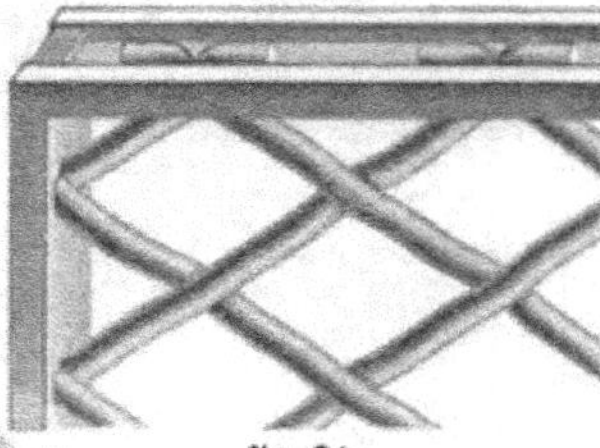

No. 21.

Cut showing method of making Nos. 4, 7, 8, 9 and 10 Crimped Wire Window Guards with a channel iron frame, and how the wire is secured to it.

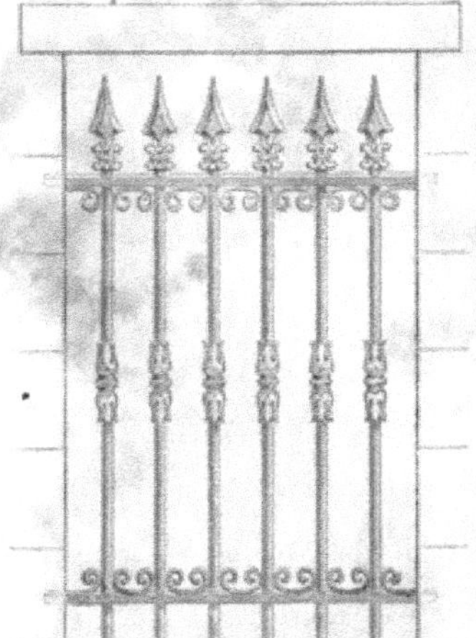

No. 11 Wrought Iron Guard.

Made of ½ inch or ⅝ round iron. Price, $6.00 to $10.00 per window, according to size.

Without ornaments.................................$4.00 to $8.00

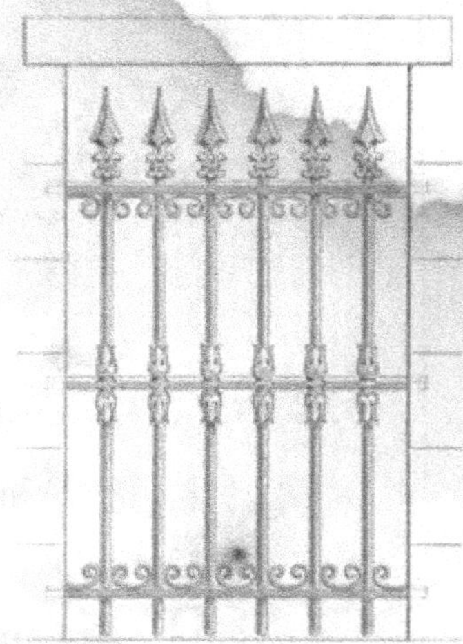

No. 12 Wrought Iron Guard.

Made of ½ inch or ⅝ round iron. Price $8.00 to $12.00 per window, according to size.

Without ornaments.................................$6.00 to $10.00

Iron Stair Plate.

Width, 6 inches; length, 18, 24, 30, 36 or 48 inches. Price, 40c. per lineal foot.

Estimates for stair work and iron work of all kinds furnished upon application.

No. 45 Iron Window Guard.

For Windows, Front Doors, Public Buildings, &c.

Prices, 75c. to $1.25 per sq. ft., according to size and quantity required.

Cast iron guards of all sizes and designs made to order from Architects' drawings price according to sizes.

No. 1½ Wrought Iron Grating.

For sidewalks in front of store windows. It is made of 1½x½ inch wrought iron, with 2x½ inch iron frame made to set into the stone flagging or wood frame work. Can be made any size and to open on hinges if desired. Price, 7½c. per lb. If to open on hinges, $2 extra on each grate. Special estimates given for large orders.

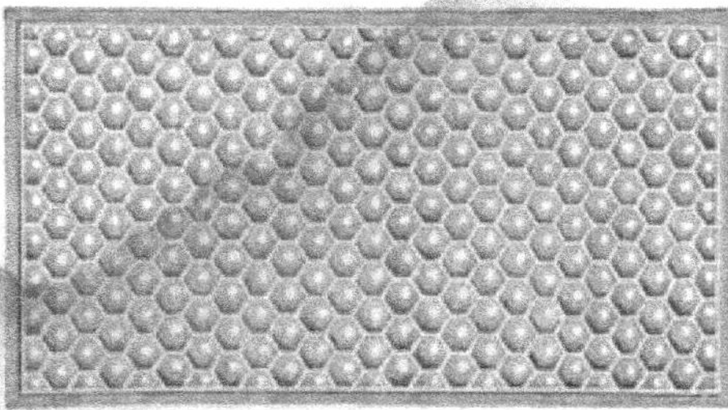

Illuminating Sidewalk Lights or Grating.

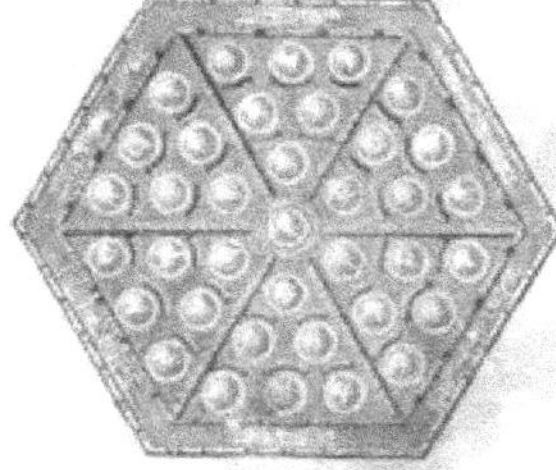

Coal Hole Grating.

This is the most desirable sidewalk grating now in use. It possesses all the strength of a wrought iron grate, admits the free passage of light, while preventing the entrance of dirt or water from the outside. Also vault and coal-hole covers of the same description. These sidewalk lights or grating for beauty of finish strength of plate, convenience of walking, firmness of tread and protection of the glass from injury, are superior to all others in the market.

Parties desiring estimates will please send a sketch showing all the dimensions. The size of the panels are 30 inches wide by 60 inches long, but they can be made to fit different size openings as may be desired; also round coal-hole or vault light and ventilators, &c.

Patent Force Pump and Garden Sprinkler.

The simplest and best Fire Extinguisher and Force Pump in the world. It has a capacity of throwing 360 gallons of water per hour, a distance of 40 feet, and will put out as much fire as any ordinary steam pump. Also, indispensable for washing windows, carriages, watering streets, sidewalks, gardens and plants, pumping out boats, &c. By putting the spray nozzle, which comes with the pump, on the end, it can be used for sprinkling fluids and liquids on trees, shrubs and plants for destroying insects. Price, $8.00 each.

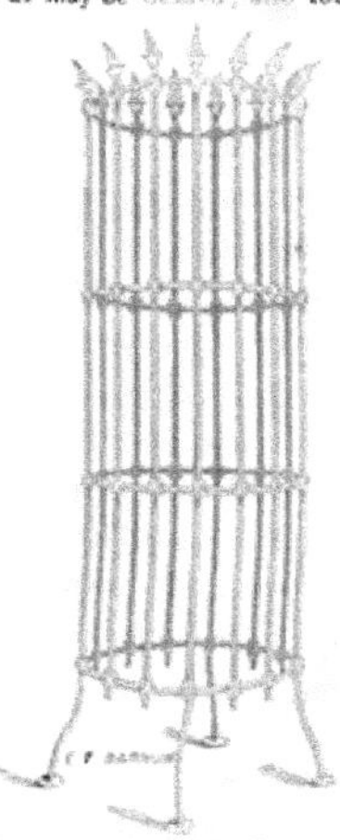

Ornamental Tree Guard.

Wrought iron, 6 feet high.................. each, $7.50
Crimped Wire, 3 inch diamond mesh..... " 6.00

They are very heavy and will last for years.

CHEESE and PROVISION SAFES.

No. 5 Square Counter Safe, $4.00.

These are intended for grocery stores, to put on counters, to hold cheese and small groceries, or for private families. It has a revolving bottom inside, and is made square, with wire cloth on each side 27 inches high and 23 inches diameter. They are also a convenient safe for family use.

No. 1, 4 shelves, 50 ins. high, 37 ins. long $8.00
" 2, 5 " 57 " 44 " 10.00

Handsomely oak grained, with figured wire cloth on the two doors, and at each end, which gives free circulation of air, and protects the meat, milk, etc. from flies and insects.

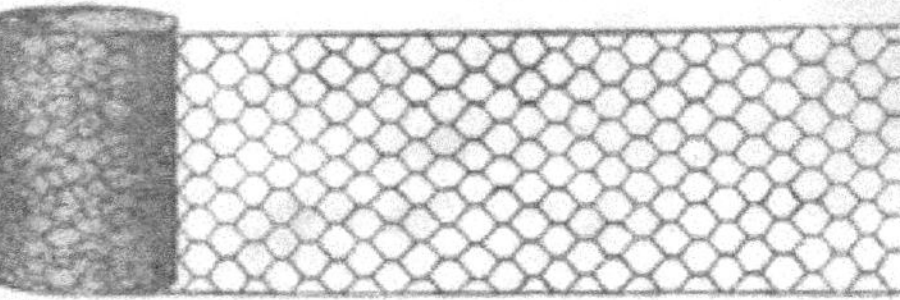

Cut showing Bale of Galvanized Netting, partly unrolled.

ROLLS CONTAIN 50 YARDS EACH.

Galvanized Wire Netting is used for a great variety of purposes where an inexpensive protection is required. It is suitable for indoors or out. It never rusts, being galvanized after making. Special prices for large orders.

Showing 3-4 Inch Mesh, exact size.

[illegible] inch mesh,	No.	20	wire	per sq. foot,	9 cts.
[illegible] "	"	22	"	" "	15 cts.
[illegible] "	"	22	"	" "	22 cts.
1 "	"	19	" for Screens, Pheasantries, Small Birds, etc.	" "	6½ cts.
1 "	"	18	" used for Glue Manufactories	" "	8½ cts.
1½ "	"	18	"	" "	4½ cts.
1½ "	"	19	"	" "	3½ cts.
2 "	"	18	" for ordinary Poultry Netting	" "	[illegible] cts.

This style of netting is made from galvanized wire, which prevents it from rusting. It is woven into a web of twisted hexagonal shaped meshes, which are twisted together in such a manner they cannot be displaced. It is used extensively in the construction of henneries, pigeon houses, rabbit hatches, aviaries, poultry hurdles, drying glue, fruit, etc. The most desirable width to sell is 36 inch. See latter part of this book for coarse heavy twisted netting, used for cattle fencing, sheep and park enclosures, etc.

BIRD CAGES.

No. 6 Robin, Mocking Bird and Breeding Cages.

Size, 16 x 9 inches .... each, $2.00	Size, 21 x 14 inches .... each, $3.00
" 18 x 11 " .... " 2.25	" 24 x 16 " .... " 3.50
" 20 x 12½ " .... " 2.50	" 30 x 16 " Mocking Bird Size .... " 4.50

These cages are well finished with walnut colored frames, moveable drawers, and are the finest wood bird cages in the market.

The different sizes can be nested for shipping.

WIRE SQUIRREL CAGE.

No. 1, for Grey Squirrel, 17x15 inches, 18 inches high. . $6.00
" 2 for Red " 14x12 " 15 " " 5.00

With wheel inside, the strongest, cheapest and best Cage made.

No. 1 Extra Heavy Parrot Cage.

No. 1, 21 in. high, 14½ in. diameter, each...... $5.00
" 2, 24 " " 17 " " " 5.50

These are made of heavy wire, and are very strong, and much more attractive than the cut represents.

Wire Rope and Sash Cord.

No. 5 PLIABLE WIRE HOISTING ROPE

IRON OR BESSEMER STEEL.

WITH 133 WIRES, 19 TO THE STRAND.

FOR ELEVATORS, HOISTING and RUNNING ROPES.

Diameter.	Breaking strain in tons of 2,000 pounds.	Proper working load in tons of 2,000 lbs.	Min. size of drum or sheave in feet.	Price per foot, in cts.
2¼	74	15	8	98
2	65	13	7	76
1¾	54	11	6½	63
1⅝	44	9	5	53
1½	35	7	4½	43
1⅜	27	5½	4	36
1¼	20	4	3½	29
1	16	3	3	23
⅞	11½	2½	2¾	19
¾	8.64	1¾	2¼	15
⅝	5.13	1¼	2	14
9/16	4.27	¾	1¾	13
½	3.48	½	1½	12

Tiller Rope.

Tiller Rope ⅜ inch diameter, 19 cents per foot.

" " ½ " 16 " "

WIRE SASH CORD OR ROPE.

PRICE PER FOOT.

No.	Diameter.	Iron.	Tinned.	Copper.
3	⅛	1½	2	3
2	5/32	2	2½	4
1	3/16	2¼	3	5
0	7/32	3¼	3¾	6½
00	¼	3¾	4½	8

This cord is used for a great many purposes where strength and durability are required. It can be used to run over small or large pulleys, but should not be kept in constant use. It is intended more for a direct strain.

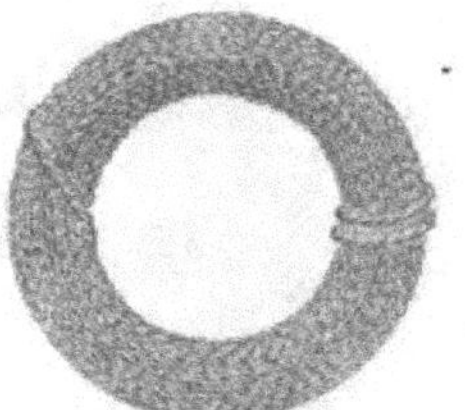

PATENT WIRE PICTURE CORD.

Price per coil of 25 yards.

Number.	Plated.	Gold.
No. 1	30 cents	$1.00
" 2	36 "	1.50
" 3	75 "	2.00

This new style wire picture cord is braided and very much more pliable, and in every way superior to the old style twisted cord, which was liable to kink, and consequently easily broken. Send a sample order and you will never use worsted, (which is easily destroyed by insects,) after trying this.

☞ Wire rope of all kinds in stock. Send for special Catalogue and price list.

STABLE FIXTURES.

Corner Oat Manger.

No. 17, each $2.75

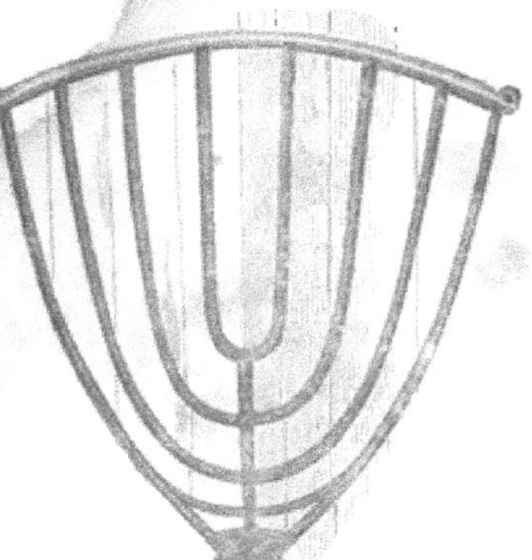

Cast Iron Corner Hay Rack.

No. 15½, each $2.75

No stable is complete without the improved Stable Fixtures. Take two of the Corner Racks together, and they make a half circle. A liberal discount by the dozen.

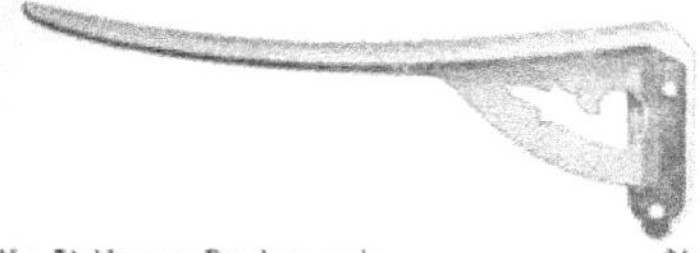

No. 71 Harness Bracket, each $1.00

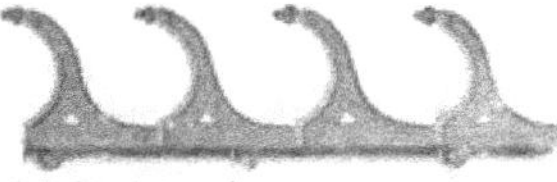

No. 74 Pole Bracket, each $1.25

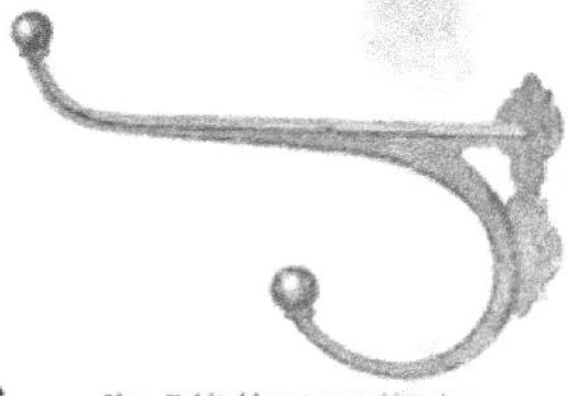

No. 74½ Harness Hooks.

6 inch, per dozen $1.50
8 " " 2.50
10½ " " 6.50

Sponge and Wet Brush Rack.

No. 66½, each $1.40

Hitching Ring.

No. 43, each........................

Blanket Brackets.

No. 20, per pair $1.50
" 20, " with wood roller 2.00

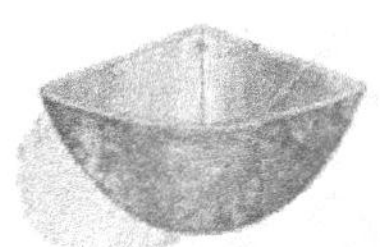

Corner Salt Dish.

No. 66, each $1.35

Cast Iron Hitching Posts.

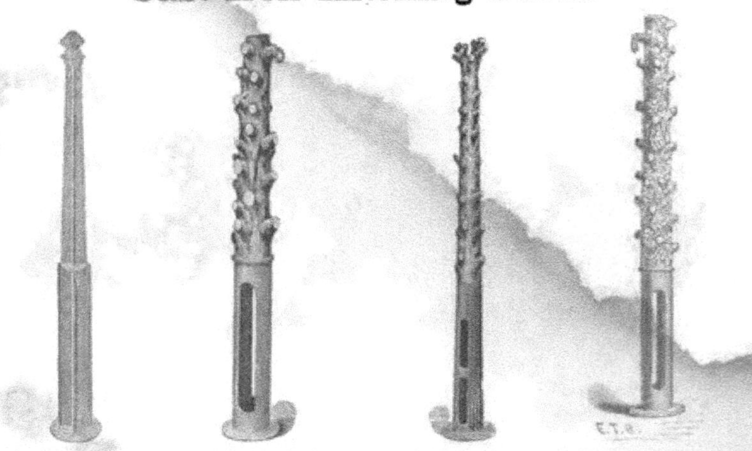

No. 5 Cast Post	No. 6 Large Rustic.	No. 7 Rustic. Imitation of Hickory.	No. 8 Grape Vine Rustic
Each $4.50	Each $6.50	Each $5.50	Each $6.50

All are made to set from 2 to 2½ feet into the ground.

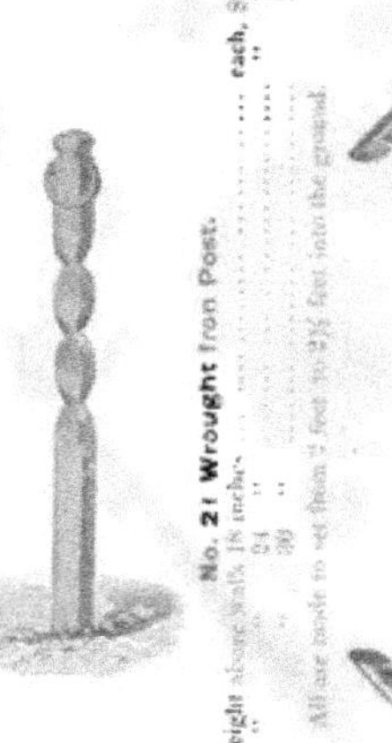

No. 21 Wrought Iron Post.

1⅝ × ⅝, height above walk 18 inches .... each, $2.00
1⅝ × ⅝, " " 24 " .... " 2.50
1⅝ × ⅝, " " 30 " .... " 3.00

All are made to set from 2 feet to 2½ feet into the ground.

Tower's Adjustable Hand-Cuffs.

Hand-Cuffs, polished .... per pair, $4.00
" extra finished and nickel plated .... " 5.00
Leg Irons, polished .... " 6.50
" plated .... " 8.00

Instantly adjustable to limbs of any size, so that they will not be found in any case either too large or too small.

Tower's Chain Nippers to clasp the wrist .... each, 1.00

Police goods of all kinds in stock. Send for special catalogue.

No. 30 Store Stool.

Green or Crimson, Plush, each $4.25
Leather cover .... " 3.50
Walnut top .... " 2.75

Special price by the dozen.

WROUGHT IRON BEDSTEADS.

No. 4 WROUGHT IRON BEDSTEAD.

2 feet 6 inches wide, 6 feet long	$5.00
3 " " 6 "	6.00

This is a very popular light Iron Bedstead, and is the best for the price in the market. It is set upon casters, and particularly adapted for using in stores, &c., on account of the peculiar manner in which they are put together, which prevents their being easily disjointed, and yet so made that they can be easily taken apart for shipping, or storing when not in use. For larger sizes and other styles send for special catalogue.

No. 1 Ornamental Wire Sign.

8 feet long, 5 feet high, black letters	each,	$25.00
10 " 6 " "	"	30.00
12 " 6½ " "	"	35.00
14 " 7 " "	"	40.00
16 " 7½ " "	"	50.00
18 " 8 " "	"	65.00

This Sign forms a beautiful finish to the roof of Banks or other buildings, and is made large and showy, according to the size ordered. Its appearance upon the roof will be much more attractive than the cut represents. Wire Signs of all sizes made to order at lowest rates. Avoid putting too much reading matter on, the less you have the larger the letters can be made and consequently are read much farther off.

Open Work Wire Banners and Signs.

No. 4 Open Wire Banner.

To Suspend in Front of Stores or Across Streets.

5 feet long	2½ feet high, black letters	each,	$16.00
6 "	3 " "	"	18.00
7 "	4 " "	"	20.00
8 "	4½ " "	"	22.00
9 "	5 " "	"	25.00
10 "	5½ " "	"	[illegible]
12 "	6 " "	"	30.00
14 "	7 " "	"	32.00
16 "	8 " "	"	35.00

Any other sizes made to order at proportionate prices. These signs are now largely used to suspend across street. For this purpose I furnish galvanized wire rope of a suitable size at from 6 to 12 cts. per foot, according to the strength required. Strangers coming into town can see this sign 8 or 10 blocks away.

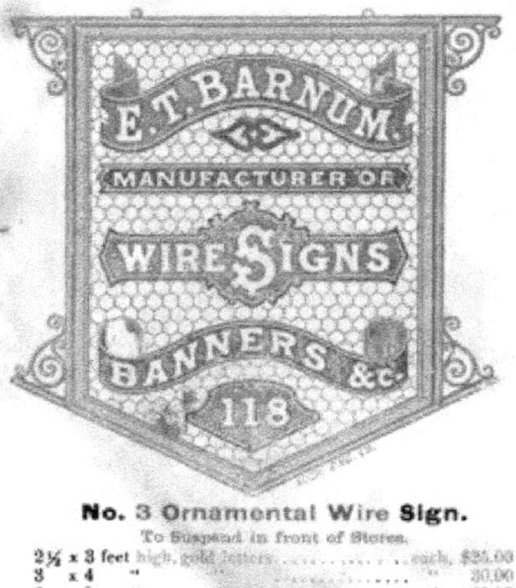

No. 3 Ornamental Wire Sign.

To Suspend in front of Stores.

2½ x 3 feet high, gold letters		each,	$25.00
3 x 4 "	"	"	30.00
4 x 5 "	"	"	35.00

This is a very popular and attractive banner. It is made entirely of metal, painted a rich bright color and relieved with gold. The letters are gilded with pure gold leaf upon a metal surface, upon both sides of the banner, so that it reads alike on either side, especially appropriate to hang out from the first story over the walk.

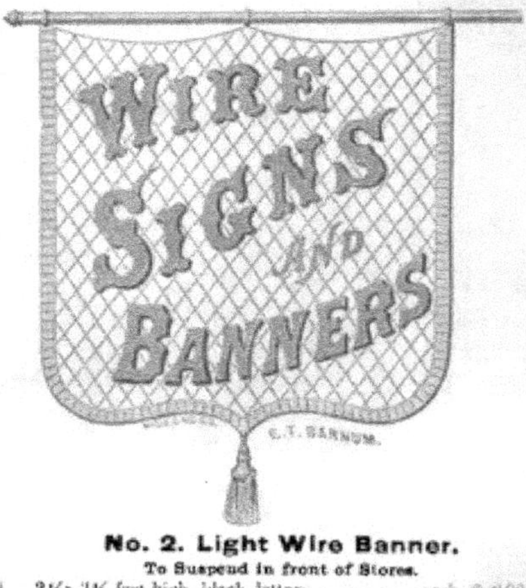

No. 2. Light Wire Banner.

To Suspend in front of Stores.

2½ x 3½ feet high, black letters		each,	$ 8.00
3 x 4 "	"	"	10.00
4 x 5 "	"	"	12.00
4½ x 6 "	"	"	15.00

This banner is much less expensive than No. 3; the letters being painted black instead of gold. They are placed upon the open wire work, and can be distinctly read upon both sides at a great distance (one side backwards). Any size not given above made to order at a proportionate price.

IRON CRESTING, or ROOF RAILING.

I am now furnishing this class of work for large buildings and private residences in all parts of the country. It is very ornamental and a great improvement to any building. If a plan of the roof is forwarded, giving the exact measurements where the cresting is to set, it will be cut and fitted so there will be no trouble in putting up. I charge for the full pieces required, and give the cresting one coat of black paint. On flat roofs it is usually set on a narrow strip of wood which is raised about two inches from the roof by being set on small blocks at a suitable distance apart. On sloping roofs or ridges this is not required, but instead, a small piece of sheet rubber is used under each foot, to prevent leaking. On flat roofs, cresting should be set about four inches from edge.

In ordering, state whether it is for a flat, sloping, or ridged roof. If for irregular shaped, sloping or ridge roofs, a small additional price will be charged for fitting.

No. 1 Cresting, 21 inches high, per lineal foot 40c
" 1 " 24 " " 45c
" 1 " 30 " " 50c

No. 2 Cresting, 18 inches high, per lineal foot 45c
" 2 " 24 " " 50c

No. 3 Cresting, 21 inches high, per lineal foot 60c
" 3 " 30 " " 70c

☞ Special prices for large orders.

No. 11 B Cresting, 15 inches high.... ....per foot, 40c

This is a good pattern for porches, piazzas, ridges, &c.

No. 20 Cresting, 15 inches high ....per foot, 40c

Ornamental Iron Finials for Cresting.

Used for Corner Posts

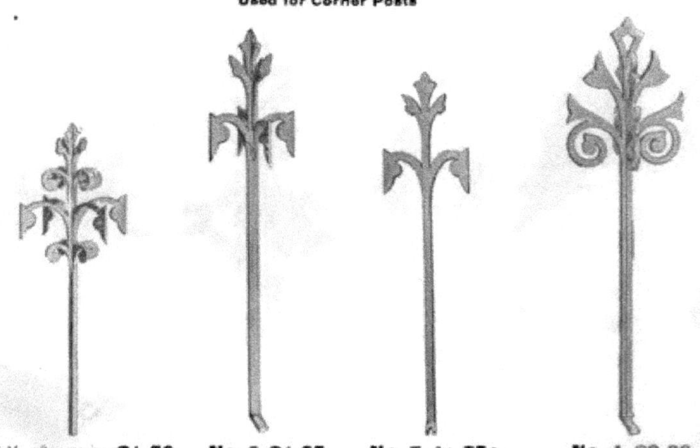

No. 1X, 18 to 30 in., **$1.50** **No. 6, $1.25** **No. 7,** flat, **75c** **No. 4, $2.00**

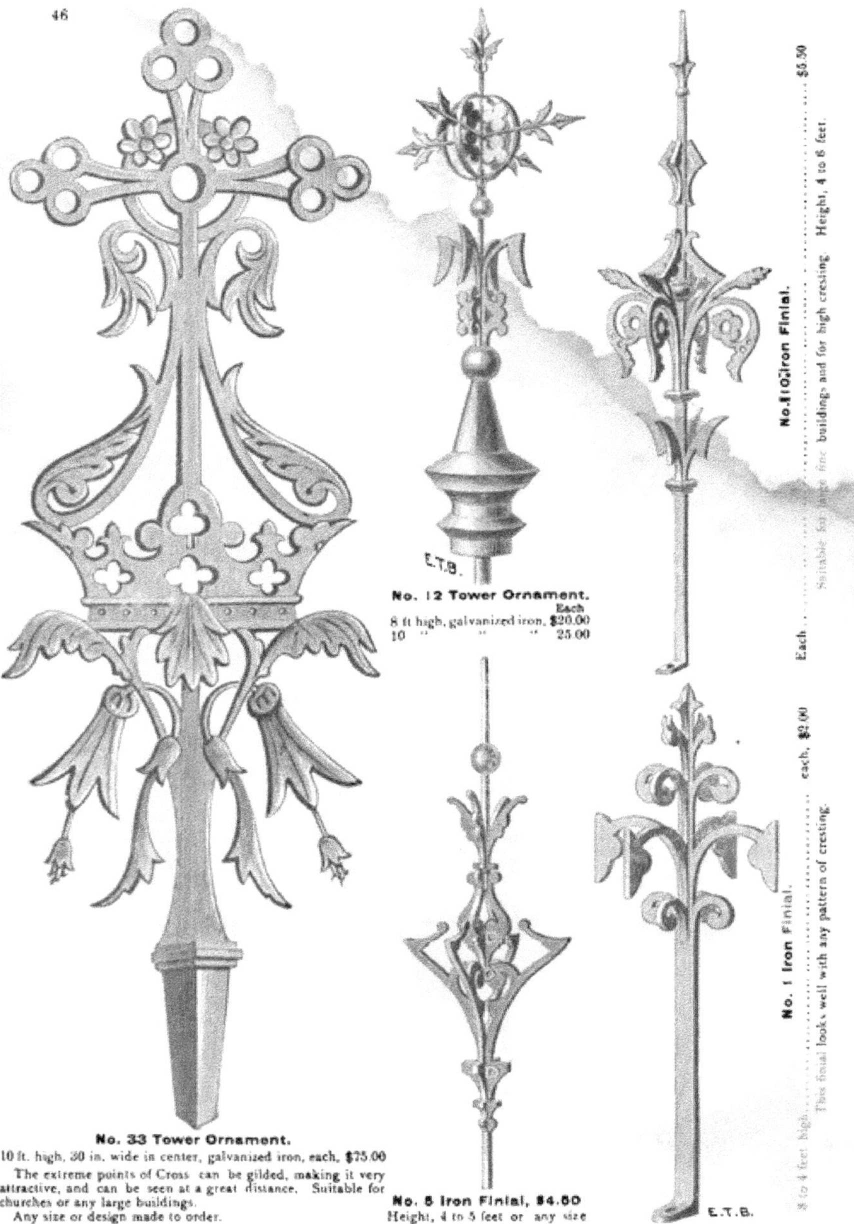

No. 12 Tower Ornament.

Each
8 ft high, galvanized iron, $20.00
10 " " " 25.00

No. 110 Iron Finial.

Each $5.50
Suitable for large fine buildings and for high cresting. Height, 4 to 6 feet.

No. 1 Iron Finial.

3 to 4 feet high .... each, $2.00
This finial looks well with any pattern of cresting.

No. 33 Tower Ornament.

10 ft. high, 30 in. wide in center, galvanized iron, each, $75.00

The extreme points of Cross can be gilded, making it very attractive, and can be seen at a great distance. Suitable for churches or any large buildings.

Any size or design made to order.

No. 5 Iron Finial, $4.50

Height, 4 to 5 feet or any size

GALVANIZED TWISTED WIRE FENCE.

(Showing Bale partly enrolled and stapled to wood Posts.)

No. 1½, 24 inches high, made from Nos. 10 and 14 wire, 4 inch mesh.......................... per lineal rod, $2.25
" 2½, 30 " " " 10 and 14 " 4 " .... " $2.50
" 3½, 36 " " " 10 and 14 " 4 " .... " 2.75

A wood strip at top and bottom is sometimes used which greatly improves the appearance as well as adds to the strength of the fence.

This fence is used for inclosing railroads, farms, lawns, sheep enclosures, etc., and for the construction of arbors, henneries, trellises and aviaries; also for protecting hedges, nurseries, shrubs, and many other uses. It is proof against prairie fires, will not cause the drifting of snow, and is not affected by high winds. It is easily put up with staples, and at any time can be taken down and removed and put up in any other place, and in this way will last for 25 years. Any mesh or width made to order.

No. 00 Pattern Light Wire Railing.

15 inches high...per lineal foot, 30c
20 " .... " 35c
24 " .... " 40c

This style of work is used in any place requiring a light and inexpensive railing, such as lawn divisions, decks of vessels, verandas, top of poultry inclosures, etc. It is made in any lengths desired, from 10 to 30 feet. If there are posts it can be fastened to them or secured with screws as shown in cut, at intervals of 8 to 10 ft. Can be made any height.

No. 1 Pattern Wire Fence and Railing.

24 inches high with cedar foundations.......................................per lineal foot, $1.10
30 " " " " .... " " 1.15
36 " " " " .... " " 1.20

For office fencing and balcony railing the feet are fastened to the floor with screws, same as shown in cut, painted brown or a wine color and bronzed. For out-door use they are usually painted green. If stone foundation posts are preferred to cedar they can be furnished, costing $1.00 extra each. Iron foundations, 75 cts. each. An extra charge will be made if more than one gate is required to every 50 feet of fence.

NO. 10 PATTERN HEAVY WIRE FENCE.

		[illegible]	[illegible]
PRICE—24 inches high, with cedar foundation posts	.... per lineal foot,	$1.00	$ .80
30 " " "	.... " "	1.10	.90
36 " " "	.... " "	1.20	1.00

This is a neat style of wire fence, and is especially intended for Cemetery Lots, Division Fences, etc. The top bar is 1 inch x 1/4 channel iron, and the wire work is made about 3 inch diamond mesh, of 3/16 to 5/16 heavy crimped wire, which is put through the channel iron at top and fastened at bottom to 3/4 inch iron, which makes a strong, handsome and durable Fence. Above prices include the cedar foundation posts for setting and one gate. Stone foundations, $1.00 each extra. Iron foundations, 75 cts. each. The latter is generally used.

☞ This style is also used for Office enclosures, and for this purpose is made to fasten to floor with screws as shown in No. 1 wire fence.

Copyrighted 1878

No. 11 PATTERN HEAVY WIRE FENCE.

FENCE—30 inches high, with iron foundation posts............per lineal foot, $1.25
36 " " " .... " 1.35
Gates with name lettered in arch as shown in cut, $3.00 extra.

This is a new and ornamental pattern with bars of 1x½ channel iron, and about 3 inch mesh heavy crimped wire work. This is a desirable pattern for front fences, cemetery lots, office enclosures, etc. It is made in sections 7 to 8 feet long and painted and finished complete ready for setting.

No. 7 PATTERN OF HEAVY WROUGHT IRON AND WIRE FENCE OR RAILING.

This is a very ornamental pattern of Wire Fence or Railing, and is stronger than many Iron Fences. It is made of [illegible] inch wire, about 3-inch mesh, in a tasteful pattern, with neat rosettes and pickets, and has a strong iron frame, which makes a fence suitable for the front of a fine residence, or office enclosure for Banks or Stores, where something handsome and durable is required. If made with plain wrought iron posts, except for gates and ends of fence, deduct 30 cents per foot from following prices:

Price—30 inches high, including iron posts, with iron foundations, per foot $2.00
36 " " " " " " 2.50

If wanted to set on stone coping, or any other purpose not requiring foundations, deduct 10 cents per foot from above prices. Gates and fastenings are $3.00 extra. The same railing made all plain diamond work (as shown at top and bottom of cut), and without rosettes, 75 cents per foot less. The style of posts will depend upon the height and style of work, and short and irregular pieces will cost extra.

E. T. BARNUM.

Copyrighted 18[illegible].

No. 27 PATTERN WROUGHT IRON FENCE WITH ORNAMENTS.

This pattern is the same as No. 29 on page 53 without the scroll ornaments. It is a very handsome style of low priced fence. The cast posts are placed each side of gates and at ends of fence, and are set into the ground from 2 to 2½ feet, according to the height of the fence. Substantial wrought iron posts are set every 7 or 8 feet as shown in No. 28 fence on the next page, the entire length of fence.

Price made with ½ inch upright pickets,	24 inches high, including posts ready for setting	....	per foot,	$1.00
" " [illegible] " "	30 " " " "	....	"	1.10
" " [illegible] " "	36 " " " "	....	"	1.15
" " [illegible] " "	30 " " " "	....	"	1.20

☞ Single gates are usually made 3 feet wide; double gates are made about 8 feet wide and are $5.00 to $10.00 extra, according to size and style.

NO. 28 PATTERN WROUGHT IRON FENCE.

This is a new pattern, open and attractive, low in price and intended for residences, cemeteries, &c. The posts at sides of gates and at ends of fence are ornamental cast iron, the balance wrought iron with bases attached as shown in cut, which set firmly into the ground, and are so made [illegible] resist any action of the frost.

This fence is made in sections from 7 to 8 feet long. Width of single gates, 3 feet. Double gates, which are [illegible] extra, about 8 feet wide.

Price, made with ½ inch upright pickets, 24 inches high, including posts as [illegible] .... per foot, $1.10
" " ½ " " 30 " " " .... " 1.20
" " ½ " " 36 " " " .... " 1.25

☞ For a cemetery lot the ornamental cast iron posts are placed at each corner of the lot and each side of the gate, thus making one of the best fences for the price now in the market.

E. T. BARNUM

Copyrighted [illegible].

No. 29 PATTERN WROUGHT IRON FENCE WITH ORNAMENTS.

This new pattern is the handsomest low priced fence now in the market. The usual height for residences is 36 inch; for cemetery lots 24 or [illegible] inch. Single gates are made 3 feet, and double gates, which are $[illegible] to $[illegible] extra, about 8 feet wide.

Price made with ½ inch upright pickets, [illegible] inches high, including posts ready for setting .... per foot, $[illegible]
" " ⅝ " [illegible] " " " " .... " [illegible]
" " ¾ " [illegible] " " " " .... " [illegible]
" " ⅞ " [illegible] " " " " .... " [illegible]

☞ The posts and foundations are all in one piece and set firmly into the ground, from 2 to 2½ feet, according to height of the fence.

No. 44 PATTERN WROUGHT IRON FENCE.

This fence is made with cast posts as shown in cut, which set firmly into the ground about 2½ feet; these are used for gates and corners or ends of fence. The intermediate posts are wrought iron with an ornamental top and a substantial iron foundation, which sets firmly into the ground (about 3¼ feet) with a bottom flange 2x4 inches, which effectually resists the action of the frost.

Price, made with [illegible] inch upright pickets,	30 inches high, complete, ready for setting	per lineal foot,	$1.40
" " [illegible] "	30 " " " " "	"	1.50
" " [illegible] "	36 " " " " "	"	1.60
" " [illegible] "	42 " " " " "	"	1.65
" " [illegible] "	42 " " " " "	"	1.85
" " [illegible] "	48 " " " " "	"	2.00

This is a good pattern for residences, park or cemetery enclosures, etc. These Fences are all new and attractive styles.
Special prices for large orders for Parks, Court Houses, Cemeteries, etc.

No. 45 PATTERN ON STONE COPING.

No. 22 Pattern Heavy Ornamental Wrought Iron Fence.

This new pattern of ornamental wrought iron fence is very attractive and intended especially for large and elegant residences. It is much preferred to cast iron, being more ornamental and not liable to be broken by any ordinary pressure. The horizontal bars are [illegible] channel iron; the pickets are [illegible] wrought iron, and all the scroll work is wrought iron, securely fastened to the pickets and bars, making this fence one of the most ornamental patterns made. This is a handsome fence to set on a stone coping, and if parties desire to furnish the foundations, or set it upon stone coping, a reasonable reduction will be made.

Price, made 36 inches high, including ornamental [illegible] posts, one gate, and iron foundations, complete, ready for setting .... per lineal foot, [illegible]
" " 42 " " " " " " .... " " [illegible]
" " 48 " " " " " " .... " " [illegible]

Single gates are made 3 feet wide, and double gates, which are extra, about 8 feet wide. Cast posts are set as shown in cut every 8 or 9 feet.
Short and irregular shaped fences or fences under 50 feet are extra in all cases.
I make a specialty of Iron Fences for fine Residences, Court Houses, Parks, &c. Special prices for large orders.

No. 32 Pattern Ornamental Carriage Gate with Side Entrance.

This style of wrought iron gate is very rich and ornamental in appearance, and intended for large, fine residences. They can be made to correspond in style with any different pattern of fence, and can be used as entrance gates for parks, cemeteries, &c.

Price of gates, made with ¾ iron pickets, 3½ or 4 feet high, without posts.......... $40.00
" " " ¾ " 3½ or 4 " including posts with foundations.......... 55.00

The width of opening for large gate is usually about 9 feet, small one 3 feet. Can make single gate on each side of double gate if desired, costing $12.00 additional. The catches, hinges and scroll work varies according to the size of gate.

No. 21 Tubular Pattern for Park or Cemetery Fences.

This is a new pattern of ornamental fence, which is suitable for enclosing public or private grounds, parks, cemetery lots, &c. It can be made with or without the chain and tassel ornaments. The bar is tubular wrought iron and the ornamental posts are cast iron and set firmly into the ground about 2½ feet, making one of the best fences of the kind now in the market. Height of post above ground 3½ feet, diameter at base 5½ inches, at center 3½ inches. Distance between bars, 18 inches and height of top bar about 2 feet 10 inches from ground. See cut of tubular gate which is used with this fence on next page.

Price, made with 2 inch outside diameter tubular bars, including posts, and complete as shown in cut, painted .... per lineal foot, $1.50
" " 2 " " " " " galvanized .... " " 1.75
" " 1½ " " " " " painted .... " " 1.25
" " 1¼ " " with plain posts, center and chain ornaments .... " " 1.00

Tubular gates with name plate to match this fence extra, each, painted, $10.00; galvanized, $13.00.

Special prices will be quoted upon application for tubular fences without posts or chain ornaments, and for the bars only with or without center and end ornaments. Also special quotations will be made for large orders of above fence for parks, &c. For this purpose it is used without the chain ornaments, and made in sections of about 10 feet each.

These Fences painted with metalic paint are much more desirable and durable than the galvanized, but [illegible] are easier kept in order, although galvanized fences can be painted at any time if so desired.

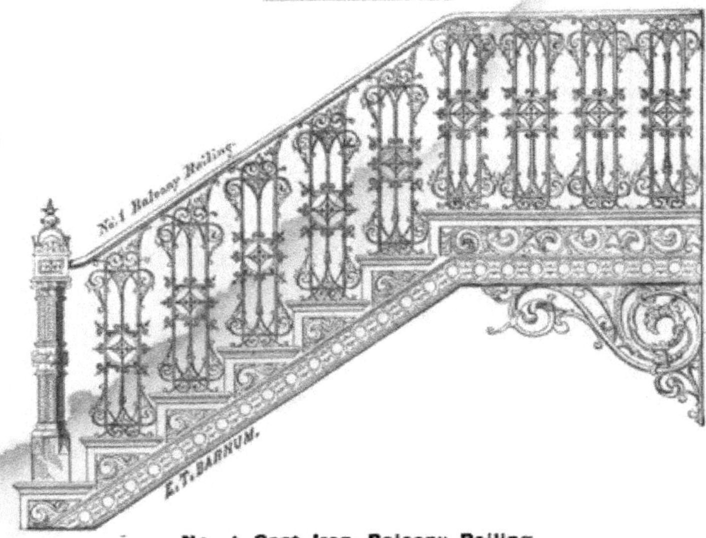

No. 1 Cast Iron Balcony Railing.

This is a new pattern and is strong and very ornamental, and when up is one of the finest balcony railings made.

Price, 30 inches high, for level balconies	per foot,	$1.50
" 30 " for stairs (not curved)	"	1.75
" 30 " plain tubular (2 rails, 1⅛ inch diameter)	"	1.00
" Cast Iron Newell Post	each,	3.50
" 30 inch Cast Brackets (for 36 inch balconies)	"	2.50
" 36 " " " 42 "	"	3.50

These brackets have wrought iron tops, which are made to fasten through a 12 inch wall.

Special estimates will be made for Cast Iron Stairs as per cut, also for Porches, Piazzas, Circular Stairs, &c. In ordering or writing for estimate send a sketch if possible, and give full particulars as to work wanted. When ordering brackets state thickness of wall.

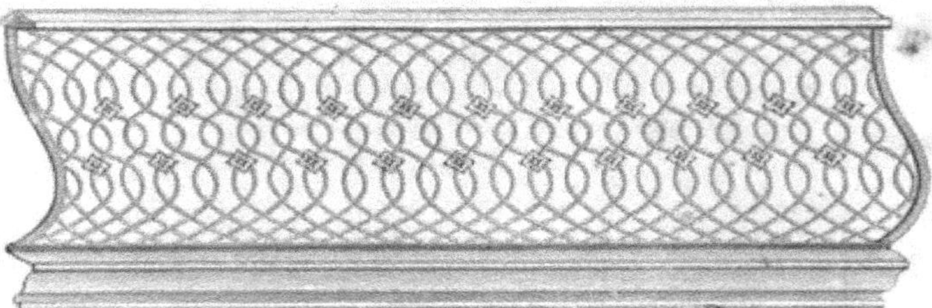

No. 7 Ornamental Curved Theatre Railing.

Made of $\frac{3}{16}$ inch heavy ornamental wire work with bronzed ornaments. It is generally used with a walnut top rail, which is covered with green or crimson plush, making a very showy and ornamental railing.

Price, 22 inches high, painted and bronzed	per foot,	$2.50
" 24 " " "	"	2.65

If wanted straight instead of curved, deduct 50 cts. per foot from above.

This work is much more showy than the cut represents. It is curved towards the front, which makes a much more convenient railing for Opera Houses, Theatres, etc., than a straight railing.

Special estimates for large orders.

Wire Bank and Office Railings.

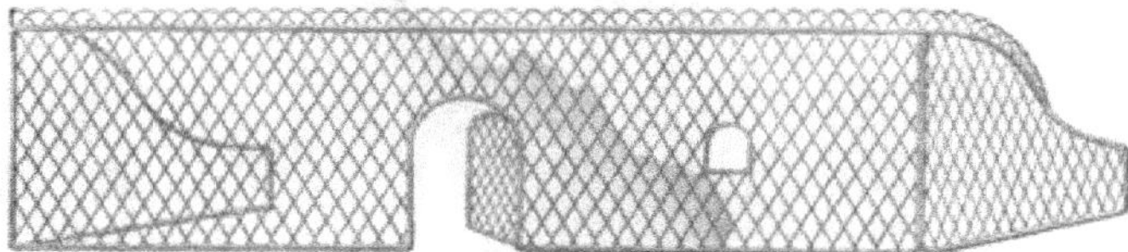

No. 1 Crimped Counter Railing.

12 inches high, or under, per running foot	...	$ .60
15 " " "	...	.75
18 " " "	...	.85
20 " " "	...	.90
24 " " "	...	1.00
30 " " "	...	1.25

Small cash hole, size 7x8 or 8x10, extra, $1.00. O. G. ends extra, each $1.00.
Cash hole with door, size 10x12 or 12x14 extra, $2.00. Without door, $1.00.

This railing is not intended for large offices, but more especially for store counters and offices where a cheap temporary railing is required. The frame is ⅝ inch round iron with 1¼ inch mesh crimped diamond wire work. It is desirable to have the ends turn in, or project over the counter to help support and strengthen it and make it firmer, unless each end can be stapled to wood-work Painted and bronzed.

No. 2½ Crimped Wire Counter Railing.

18 inches high, 1¼ inch mesh, No. 13 wire	...	per lineal foot,	$1.25
25 " 1¼ " " 13 "	...	"	1.50
30 " 1¼ " " 13 "	...	"	1.75

The price includes ornaments on top, all complete, painted and bronzed. For price of cash holes, doors and O. G. ends, see No. 1 railing.

This is a cheap railing, rather more ornamental than No. 1, but intended more especially for small stores and offices. Crimped wire at the bottom is clinched around a ⅝ inch round iron bar; the top bar is 1⅛x¼. Can be screwed to the counter, but is not self-supporting like Nos. 5 or 6 railing. Like No. 1, it is desirable to have the ends project over the counter to help support and strengthen it, unless each end can be fastened to wood-work.

☞ In ordering send a sketch showing the plan of the railing and be particular to give all the dimensions correctly; location of cash holes, sizes of the same, etc.

☞ Order all goods in this catalogue by *number and name* and not by the page.

No. 6X Wire Railing, showing Entrance Door to Private Office.

The price of this Door and Railing is the same as No. 6 Pattern, 36 inch, (see page 65) with $5.00 additional for lock and lettering as shown in cut. This style of work affords the safest protection for any Banking Office, and is preferred to glass or other enclosures on account of its strength as well as ornamental appearance.

I submit herewith a few of the many unsolicited letters which I have received from parties who have recently purchased my Wire Railings, showing how well they are appreciated.

Office of ROGERS & STUART, OWOSSO, MICH., January 29th, 1878.

MR. E. T. BARNUM, Detroit, Mich.

Dear Sir—The Wire Counter Railing we ordered of you came duly to hand and we were agreeably surprised. The job was perfect in every particular and the office presents a fine appearance. We prefer a wire railing to glass or any other, it being neat, tasty and perfectly proof against depredators. Allow us to say that you now have a standing advertisement here, and if others should want anything of this kind in the future rest assured you will be remembered.

Very truly yours,

ROGERS & STUART.

CLARKSVILLE, TEXAS, March 22, 1878.

E. T. BARNUM, Esq.

Dear Sir—The Wire Counter Railing and Wire Fence I bought of you for the Railroad Co. Bank of this place has given entire satisfaction, and is admired by those who see it. I think it just the thing, and is a much safer protection for money than glass.

Yours respectfully,

E. H. BRITTAN.

Office of FREDERICK HASS, ROCK ISLAND, ILL., May 15, 1878.

E. T. BARNUM, Esq., Detroit, Mich.

Dear Sir—I take pleasure in stating that in every instance where I have used your Counter Railing, it has given entire satisfaction, and is, in my opinion, when used in banks, offices, etc., much more preferable than glass, as it is stronger and safer, and a much better protection to valuables.

I remain yours respectfully,

FREDERICK HASS.

OSAGE NATIONAL BANK, OSAGE, IOWA, Nov. 15, 1878.

E. T. BARNUM, Esq., Detroit, Mich.

Dear Sir—I enclose Draft for $92.00, in payment for Counter Railing which is in place and fits exactly. It is a very fine piece of work, and we are well satisfied with it and much obliged.

Yours respectfully,

J. H. BRUSH, *Pres.*

SULLIVAN, ILL., November 4, 1878.

E. T. BARNUM.

Sir—Your Fence I have received and am well pleased with it. The party that received that fence, says, it is one of the handsomest fences in this county, and are well satisfied with it. I will soon get some more orders.

Truly yours,

FRED SONA.

NATIONAL BRANCH BANK OF MADISON, MADISON, IND., Nov. 26, 1878.

E. T. BARNUM, Esq., Detroit, Mich.

Dear Sir—Inclosed find check on New York, $231.74, payment for your bill enclosed for Bank Railing. You made what we wanted, and made it well. We shall see in a few days about another order for your goods.

Yours truly,

D. G. PHILLIPS, *Cashier.*

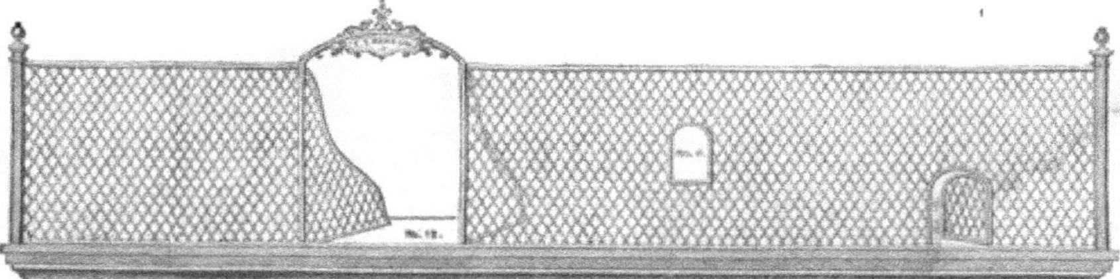

No. 3 PATTERN WIRE COUNTER RAILING.

This railing is a plain pattern, does not obstruct the light and air like glass and clumsy iron railings, and when painted and bronzed with gold it is very neat and attractive. The frame is made of iron with a bottom bar 1x¾ inch, which is fastened to the counter with screws; into this frame is crimped 1¼ inch mesh heavy diamond wire work, made out of No. 12 wire.

PRICES - 24 inches high	per square foot,	$1.00
30 "	"	.90
36 "	"	.85
42 "	"	.80
No. 13, O. G. wings or projections	each,	2.50
" 12, Arch (with lettering) extra		8.00
" 11, Cash Hole or Posts		1.30 to 2.00
" 10, Cash Hole and Door		4.00

☞ In ordering state width wanted of No. 12 opening, length of the O. G. turns each side and lettering wanted in arch; also distance wanted from bottom of railing to bottom of No. 11 opening, and sizes of openings preferred. Nos. 10 and 11 openings are made 7x8, 16x12 or 22x14.

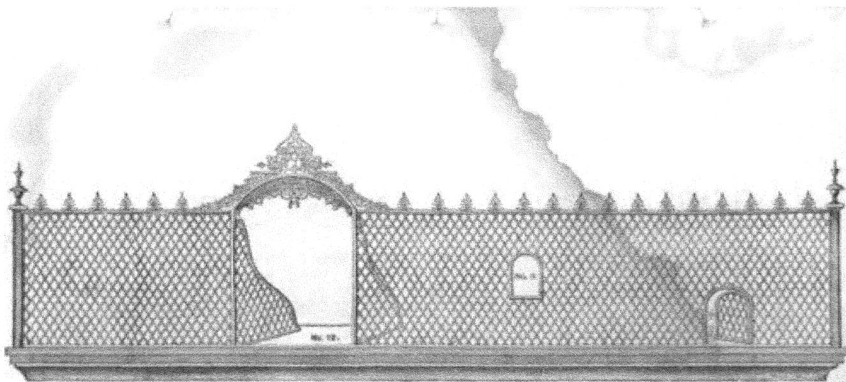

No. 4 PATTERN WIRE BANK COUNTER RAILING.

A new and desirable style for Banks, Stores, Insurance Offices, etc. Is made with a substantial iron frame which fastens securely to the counter with screws. The wire work is about 1½ inch mesh, made of No. 12 crimped wire and is very strong. All my railings are handsomely painted and finished with gold bronze.

PRICE—24 inches high ... per square foot	$1.00	No. 12, O. G. wings or projections ... each	$2.50
30 " ... " "	.90	" 12, Arch (with lettering) extra ...	3.00
36 " ... " "	.85	" 11, Cash Hole or Posts ...	1.50 to 2.00
42 " ... " "	.80	" 10, Cash Hole and Door ...	4.00

☞ In ordering state width wanted of No. 12 opening, length of the O. G. turns each side, and lettering wanted in arch; also distance wanted from bottom of railing to bottom of No. 11 opening, and sizes of openings preferred. Nos. 10 and 11 openings are made 7x8, 10x12 or 12x14.

No. 5 WIRE BANK COUNTER RAILING.

This pattern is the same as No. 4 with the addition of the bronzed ornaments in the center, which gives it a finished and attractive appearance. It can be made any shape, with or without posts, cash holes or openings. The frame is made of iron, which fastens securely to the counter with screws. This railing is handsomely painted and finished with gold bronze, and is an ornament to any office.

Prices—24 inches high	per square foot,	$1.00
30 "	"	.90
36 "	"	.85
42 "	"	.80
No. 12, O. G. wings or projections	each,	2.50
Pickets on top, extra	per lineal foot,	$ .20
Rosettes in center, same	"	.20
No. 12, Arch (with lettering) extra		3.00
" 11, Cash Hole or Base		1.50 to 2.00
" 10, Cash Hole and Base		4.00

I make a specialty of wire railings, and can turn out better styles and quality of work than any other house for the same money. Parties when ordering will please bear this in mind when comparing prices, as many are copying my cuts and then claim to make the same quality of work.

☞ In ordering state width wanted of No. 12 opening, length of the O. G. turns each side, and lettering wanted in arch; also distance wanted from bottom of railing to bottom of No. 11 opening, and sizes of openings preferred. Nos. 10 and 11 openings are made 7x8, 10x12 or 12x14.

No. 6 WIRE BANK AND OFFICE RAILING.

PRICES.

24 inches high, per square foot	$1.20	Pickets on top extra, per lineal foot	.20
30 " "	1.45	Rosettes in center extra, "	.25
36 " "	1.90	Posts, or No. 10 and 21 Cash Holes, each	$1.00 to $2.00
No. 13 Cash Hole and Door, extra	4.00	No. 14 Cash Hole, with glass shelf, arch and lettering	8.00

☞ In ordering state distance wanted from bottom of railing to bottom of Nos. 14 and 15 openings. The sizes of the openings are as follows: Nos. 13 and 21, 7x8, 10x12 or 12x14 high; No. 14, 15x15 high; No. 15, 7x8 or 10x12 high. I am now using an ornamental frame for all cash holes, which greatly adds to the appearance of the railing. My wire railings are fast taking the place of glass, as no bank officer will leave valuables on a counter with simply a glass partition between them and a bold thief. Having fitted up several hundred banking houses throughout the country, I feel confident that no argument is necessary to convince any one that my railings are indispensable and superior to any other railing [illegible]. I have many [illegible] testimonials from a large number of those receiving my work, which will bear me out in this statement, a few of which I publish on page 61.

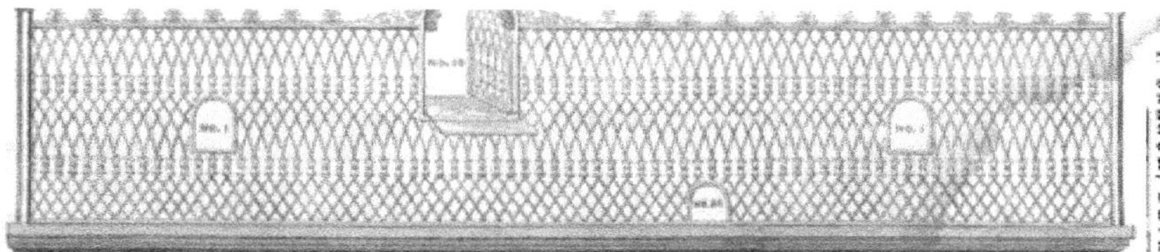

No. 7 WIRE BANK AND OFFICE RAILING.

PRICES.

[illegible] high, per square foot	$1.25	Nos. 1 and [illegible] Cash Hole, [illegible] inches, each	$1.50 to $2.00
[illegible]	1.20	Rosettes [illegible]	.75
No. [illegible] Hole, 14x18 inch, with glass shelf, door and lettering	10.00	Posts, [illegible]	$1.50 to $2.00

[illegible] In ordering state the distance from bottom of railing to bottom of Nos. 1 and 18 [illegible] sizes preferred for Nos. 1 and 20, and lettering for each. No. 18 opening is made one size, 14x18 high.

This [illegible] double gothic [illegible] is similar to No. 6, but with a double row of [illegible] with the bronzed ornaments on top gives it a very [illegible] appearance. It is finished in gold bronze and is very attractive. It is [illegible] free offices, and can be made with such cash holes, [illegible] doors as [illegible] to suit the convenience of the purchaser.

No. 8 HEAVY WIRE BANK COUNTER RAILING.

This new pattern is one of the handsomest and most attractive railings made, and is especially adapted for banks and large, fine offices. By a new process the wires are gracefully put together and worked into an ornamental pattern with handsome bronzed ornaments, which gives it a finished and attractive appearance. This railing is painted and finished with gold bronze, and is acknowledged to be one of the finest railings ever introduced. The frame is made of iron with a bottom bar, which is fastened to the counter with screws. The diamond meshes are about 1⅝ to 2 inch heavy wire work, made out of No. 10 wire.

PRICES.

30 inches high .... per square foot,	$1.40	No. 19 opening, 7x8, 10x12 or 12x14, without door, extra	[illegible]
36 " .... "	1.35	Ornamental posts .... each,	[illegible]
42 " .... "	1.30	Arch with lettering (over No. 17 Gates) .... "	[illegible]
48 " .... "	1.25	Bronzed rosettes and pickets (3 rows) .... per lineal foot,	[illegible]
No. 18 opening, 7x8, 10x12 or 12x14, with door, extra	4.00	No. 17 double gate, 3 feet wide, with fastenings .... each,	[illegible]

This new style, like all other patterns, can be made to order any size or shape desired with such styles of cash holes, openings, etc., as the business of the office may require. In ordering mention the sizes preferred for openings and the lettering desired for No. 17 gates.

No. 9 HEAVY WIRE BANK COUTNER RAILING.

A new pattern and very ornamental. Is made of heavy crimped wire work with iron frame, and handsomely painted and **finished with gold bronze.**

24 inches high .... per square foot, $1.00	Ornamental posts .... each, 2.00
30 " 1.25	Arch with lettering over No. 15 [illegible] .... " 6.00
36 " 1.25	Bronzed [illegible] and pickets .... per lineal foot, 1.00
No. 15 opening, 7x8, 10x12 or 12x14, with door, each 4.00	No. 17 double gate, 3 feet wide, with lettering .... each, 5.00
No. 10 opening, 7x8, 10x12 or 12x14, without door, each 2.00	Counters or irregular shaped pieces extra.

I publish herewith names of a few of the many banking and business houses for whom I have recently furnished wire railings, **to whom I refer for the quality of my work.**

[illegible]

[illegible]

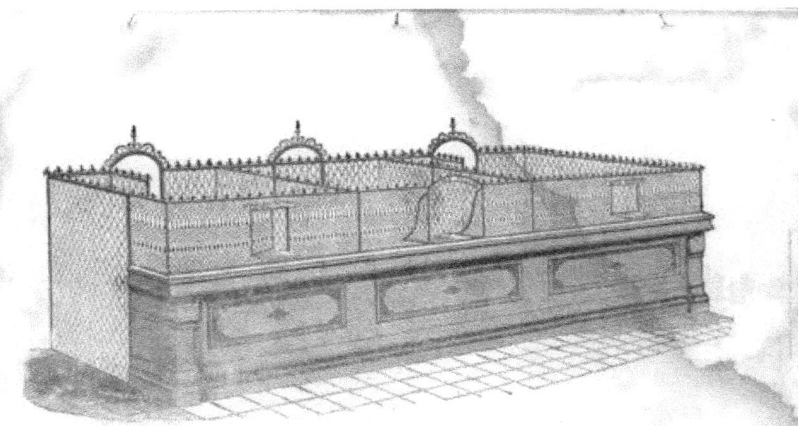

Interior View of a Banking Office, **fitted up with Wire Railing and Wire** Partitions.

This cut shows how an office may be fitted entirely with my wire railings and wire work partitions arranged for the cashier, teller and clerk's desk, having every convenience and protection against those sudden attacks from robbers to which all tellers and cashiers are exposed while in the discharge of their duties. The work may be made of any of the foregoing patterns, and provided with such cash holes, doors and openings as are best suited to the requirements of the office.

☞ I make a specialty of Bank Railings, and will furnish upon application unsolicited testimonials from some of the largest banking institutions in the country which are using my Railings, showing the satisfaction my work is giving. A few of these testimonials I publish on page 61.

Wire Cloth Signs for Store or other Windows.

LETTERED UPON **FINE PAINTED WIRE CLOTH.**

Showing No. 23 Style of Lettering upon Green Painted Wire Cloth.

Used for Signs in store or other windows. The frame is imitation of walnut and the wire cloth is fine mesh, painted green or drab, upon which any lettering desired is painted in a neat and tasteful manner, closely representing gold in color.

Prices of No. 23 Style Lettered Window Screens.

Green or Drab Cloth, frames imitation walnut	per square foot,	20c
" " " black "	" "	23c
Lettering as above, 6 inch letters	per lineal foot,	25c
" " 8 "	" "	30c
" " 10 "	" "	35c
Bordering as above	" "	10c

If gold lettering is desired see prices upon next page.

Showing No. 21 Style of Lettering upon Figured Painted Wire Cloth.

For prices add 5 cents per square foot to those given above for No. 23 style.

Painted lettering is the same as above. Gold lettering same as given upon next page.

In ordering be particular to mention which is the height, otherwise the lettering may be wrong.

If desired, large frames can be made and cloth tacked on by parties ordering the screens. For dwelling house windows, my screens can be made to fasten to the stop in a manner, so as to allow them to be raised and lowered with ease, enabling one to open and close the windows and blinds without removing the screen or window, or marring the casing.

Dwelling house window screens and doors made to order. Send for estimates.

Wire Cloth Sign for Bank or other Windows.

Lettered upon Fine Landscape Painted Wire Cloth.

Copyrighted 1879.

Showing No. 25 Style of Lettering upon Landscape Painted Wire Cloth.

The fine wire cloth of which these new and attractive screens or signs are made is coated with a ground color of light drab, and afterwards beautifully painted and decorated by hand, in imitation of mountain, water, rustic and other natural scenery, making a very handsome and useful screen and sign combined. A peculiarity of these screens, apart from their great beauty is that persons inside of a room can look out without difficulty, while those from the street cannot look in, and you are thus secluded from the gaze of outsiders. This alone is of great importance to banking institutions, when they have papers, money or valuables exposed. These screens are painted in different views, according to size and height.

PRICES.

24 to 30 inches wide, frames imitation walnut	per square foot,	$ [illegible]
24 to 30 " " black walnut	"	[illegible]
With State Arms on Landscape	extra,	[illegible]
With State Capitol or U. S. Capitol	"	[illegible]
With Store or any public building	"	
¾ inch Gold Border, around outside	per running fo[illegible]	
6 " Rich Gold Letters shaded	"	
8 " " " "		
9 " " " "		
10 " " " "		

The size of the letters will depend upon the size of the screen, and painted lette[illegible] the price of gold letters.

Any firm name or sign can be lettered upon the above screen in large, rich g[illegible] generally used with making them very attractive and indispensable, especially in banking offi[illegible]howy and ornamental [illegible]ticular and give the dimensions and say which is the height, otherwise your [illegible]
Pemp. If the windows are large the frames are usually made by parties .... per foot, $2.50
" [illegible]ip better without frame, and are easily tacked on after being recei[illegible] .... " 2.65

It special design in landscape can be made to order, such as U. S[illegible]ove.
[illegible]of Arms, Business Signs, Trade Marks, Houses, Buildings, or ar[illegible]curved towards the front, which make[illegible] Insurance Offices, Billiard Rooms, Commission Houses, etc[illegible] etc., than a straight railing.
Sp[illegible]

Made of $\frac{3}{16}$ inc
a walnut top rail, w
railing.

Price, 22 inches high, pain ... embracing three stores, on Woodward Avenue, running back 100 feet on Atwater Street
" 24 " ... Avenue, one block from the Ferry.

If wanted straight instea ... and largely increased the facilities, and are now prepared to execute all ... which
This work is much more ... NCING, &c., with ... ness and at lowest rates.
makes a much more convenie ... ame and notify the ... will please remit all or a part of
Special estimates for large orc

E. T. BARNUM.

27, 29 and 31 Woodward Avenue, Detroit, M

Please hand this to some Architect or Builder.

www.ingramcontent.com/pod-product-compliance
Lightning Source LLC
Chambersburg PA
CBHW022010190326
41519CB00010B/1462